Biogeochemistry of an oak-woodland ecosystem in the Netherlands affected by acid atmospheric deposition

Agricultural Research Report 930

N. van Breemen, W.F.J. Visser & Th. Pape (eds)

Biogeochemistry of an oak-woodland ecosystem in the Netherlands affected by acid atmospheric deposition

Pudoc Wageningen 1988

CIP-DATA-KONINKLIJKE BIBLIOTHEEK, DEN HAAG

ISBN 90 220 0916 5
NUGI 825

Printed in the Netherlands

Abstract

This book describes the results of three years (1981-1984) of biogeochemical monitoring of four plots in a small oak woodland, the 'Oude Maat', in the east-central part of the Netherlands. The original oak coppice was cut for the last time in 1939, and has been largely left to itself since then. On account of a unique variation in soil composition, with both calcareous and strongly acidic soils within a short distance in the same forested area, the woodland is particularly interesting for scientific study. From 1956 to 1964, a wealth of data on vegetation and soil fauna was collected by researchers from the former Institute for Applied Biological Research in Nature (ITBON, Arnhem). From 1981 to 1987, the fluxes of major elements from the atmosphere and through the soil at various depths were studied by scientists and students from the Agricultural University, Wageningen. The book describes how a combination of hydrologic monitoring and modelling and of soil solutes monitoring was used to arrive at a chemical balance for the soils in question, and discusses the results.

Soil composition is strongly influenced by atmospheric inputs of N from ammonium sulphate, amounting to 50-60 kg/ha.year. Strong nitrification, even at low pH, is responsible for enhanced soil acidification with soil solutions dominated by aluminium nitrate. The dominance of nitrate suggests that the ecosystems are close to 'saturation' with N. Saturation with N seems to be reached at one site, where biomass production is much lower than elsewhere, and where the ecosystem apparently is unable to absorb the incoming atmospheric N.

The monitoring results have been summarized in tabular and graphical form, and the complete data are included in microfiche form.

Contents

Part 2. Monitoring programme

Preface

This book summarizes the results of three years (1981-1984) of biogeochemical monitoring of an oak-woodland ecosystem on acidic and calcareous, sandy to loamy soils in the east-central part of the Netherlands. The project started in 1980 with the aim of estimating the fluxes of chemical components in such a system as a function of the amount of calcium carbonate in the root zone of the vegetation. G. Minderman from the Institute for Nature Management (RIN) suggested the 'Oude Maat' woodland near to the Hackfort Estate, south of Zutphen as a suitable area for the research.

Data about changes in fungal flora between 1957/1958 and 1982/1984 were derived from the Biological Station, Wijster, Communication 308, and from the Department of Sylviculture and Forest Ecology, Agricultural University, Wageningen (NL), Communication D87-01.

After partial funding by the Netherlands Foundation for Pure Research (ZWO) in 1980 and 1982, additional funds from the European Community (contract ENV 650-NL) provided for continuation of the monitoring work up to March 1984 and, in cooperation with Institute for Nature Management, for an extension of monitoring to other ecosystems (Scotch pine and heather on sandy soils).

The contributing authors are scientists of the Wageningen Agricultural University:
N. van Breemen of the Department of Soil Science and Geology;
A. Ehrenburg of the Department of Vegetation Science, Plant Ecology and Weed Science;
J.J.M. van Grinsven of the Department of Soil Science and Geology;
A.E. Jansen of the Biological Station in Wijster;
A.G. Jongmans of the Department of Soil Science and Geology;
J.P. Kools of the Department of Vegetation Science, Plant Ecology and Weed Science;
Th. Pape of the Department of Soil Science and Geology;
E.J. Velthorst of the Department of Soil Science and Geology;
W.J.F. Visser of the Department of Soil Science and Geology;
P.A.B. de Visser of the Department of Soil Science and Geology;
B.W.L. de Vries of the Biological Station in Wijster.

Many individuals who do not appear as authors of parts of this volume contributed significantly to the study. S. Slager and P.A. Burrough provided invaluable help setting up the monitoring work. We thank R.B. Heringa and B.F. Wijlens of the State Forestry Service, responsible for managing the 'Oude Maat', for providing assistance and advice whenever asked for. C. Baars of the Dutch Nature Conservancy Board assisted with some of the monitoring work. Many individuals of various departments of the Agricultural University cooperated in the project: C. Dirksen and W.J. Ackerman of the Department of Soil Science and Plant Nutrition with chemical analysis of plant material; and E. Ott and H. van Oeveren of the Department of Plant Ecology with describing the ground vegetation and measuring litter fall. At the Department of Soil Science and Geology, micromorphological descriptions were made by A. van Dis, and a large part of chemical analysis was done by A. Baars-van Osch, L.Th. Begheijn, H. Sliepenbeek, F.J. Lettink, N. Nakken-Brameyer, M.T.M.H. Lubbers and A. van Osch. Most of the typing was done by Mrs M.H. van Eldik-van Miltenburg, and drawings were prepared by G. Buurman and P.G.M. Versteeg.

Introduction

The soils of the 'Oude Maat', situated in the eastern part of the Netherlands (Fig. 1), vary from calcareous throughout, to non-calcareous down to at least 2.5 m. These varied soil conditions attracted the attention of researchers from the former Institute for Applied Biological Research in Nature (ITBON), and a wealth of data was available from earlier studies on vegetation and soil fauna, mainly from 1956 to 1964 (Minderman, 1981).

Soon after the start of chemical monitoring, it became clear that atmospheric deposition of sulphur and nitrogen dominated the biochemistry of the systems studied (van Breemen et al., 1982), and the emphasis of the research was shifted to the impact of atmospheric deposition on the ecosystem. Detailed results of this research have been reported in various articles (van Breemen, N. et al., 1982; 1984, 1986, 1987; van Breemen & Jordens, 1983; Mulder et al., 1987) but an overall picture of the situation at the 'Oude Maat', including an outline of the methods used, has been

Figure 1. Location of the study area within the Netherlands.

lacking. This book provides this overall picture, and gives background data on soil and vegetation of the research sites. Vegetational changes in vascular plants and fungi since the first surveys by ITBON workers are presented here too. The appendices summarize most of the monitoring data, for 1981-1984, while complete monitoring data for that period are given in microfiche form. Chemical and hydrological monotoring has been continued until April 1987. Results of the last three years of monitoring will be published later.

Part 1
Description of physical environment and vegetation in the ‘Oude Maat’

1. Physical environment

1.1 General description of the area *(N. van Breemen)*

The study was done in a 3.2 ha woodland in an area where the land was used mainly (75 %) for intensive grass production for stall-fed dairy cows, with here and there units for pig, chicken and egg production. The remaining 25 % of the land was under forest, generally in plots of less than 1 ha to several tens of hectares. Figure 2 gives a view of the landscape of the area. The region is flat and part of the alluvial plain of the River IJssel, a distributary of the River Rhine. The 'Oude Maat'

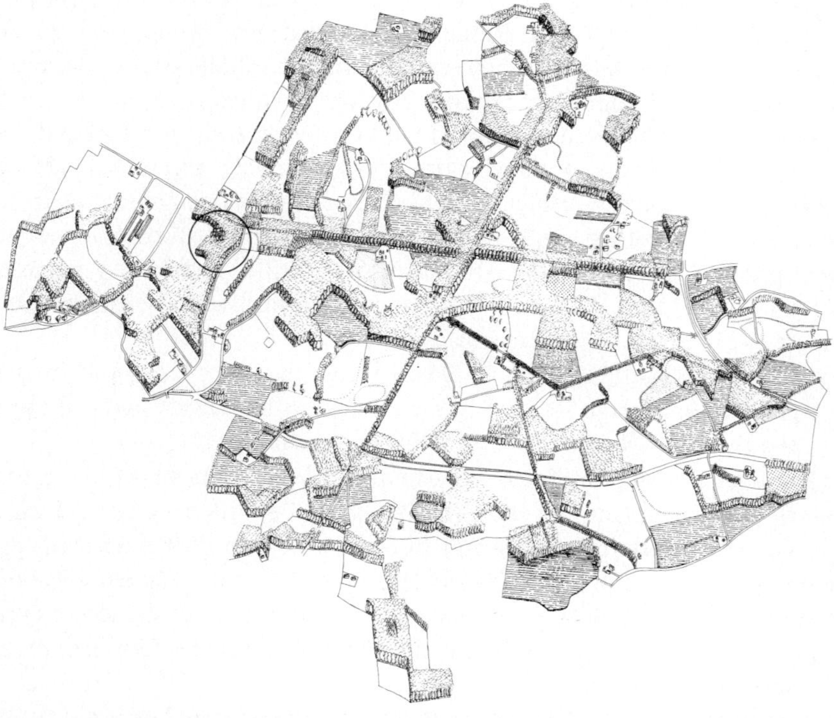

Figure 2. Bird-eye view of the landscape of the area, scale approx. 1:18000. The 'Oude Maat' is located in the circle. Reproduced by permission. (Source: W.J.C. Hoefnagel, 1985. 'Hackfort, Landschapsbeeld en cultuurhistorie' Coal, Publ. nr. 16, Research Institute for Forestry and Landscape Planning, Wageningen, The Netherlands).

may always have been under trees; old maps indicate that it has not been used as arable land since at least the 18th Century (Minderman, 1981). Human influence, however, has been intense: the forest litter has, no doubt, been removed regularly to provide bedding for cattle in the 'pot' stalls (Pape, 1972). In places, the soil has been reworked to about 1 m depth and may have been partially removed (Section 2.2). Moreover, wood was probably harvested regularly. The 'Oude Maat' used to be a coppice of oak and birch, and was cut for the last time in 1939. Apart from air pollution inputs, the area has been left largely undisturbed since then, with human interference mainly confined to the effects of improvement of land drainage in the general area in winter, and, in the 'Oude Maat' itself, to research activities (1956-1964; 1980-present). The woodland is situated in Map 33 H, Zutphen (scale 1:25,000), at coordinates 215.2/457.53. Within the 'Oude Maat', an area of 50 m × 200 m has been provided with a reference grid of 10 m x 10 m by means of wooden posts, installed in the late 1950s and renewed in the late 1970s (Fig. 7).

1.2 Geology and physiography *(A.G. Jongmans & W.J.F. Visser)*

The study area and its surroundings consist of fluviatile sediments (Kreftenheye Formation), deposited by the Pleistocene Rhine until the Middle Weichselian (Van de Meene, 1979). The Pleistocene Rhine was a braided river, with low-lying, continuous gully systems and relatively high easily erodible channel banks. Because of a highly irregular water supply, the gullies migrated frequently (Pannekoek, 1973). Differences in height between gullies and banks were generally 0.5-1 m and did not exceed 2 m. The sediments consisted mainly of (presumably calcareous) sand to loamy sand with little gravel. During the final phase of the braided river system, a thin layer of clayey material was deposited on top of the sandy sediments, especially in the lower parts of the river plain. Local dispersion of the sands by wind caused the formation of river dunes. They consist of sand with little loam. The dunes have been built up to 1-2 m above the river bank.

During the Holocene Period, fluviatile clay from the River IJssel and from some small streams coming from the east was deposited in the lowest parts of the area. This is called young river clay. Peat was formed locally.

Figure 3 shows the distribution of aeolian and alluvial sediments in the region. To the east of the river plain, cover sands have been deposited by wind during the Pleistocene. The rather flat macrorelief of the Pleistocene river-plain was first accentuated by the formation of dunes and later levelled off by sedimentation of clay and formation of peat. Still later, man accentuated some higher positions by practising plaggen-manuring (Pape, 1972) and elsewhere obliterated differences in microrelief by levelling of the land.

The hydrology of the braided river-plain is influenced by its low position. Water is supplied by seepage from the higher areas to the west (Veluwe Massif) and to the east (coversand area), and via the many east-west flowing streams. Seepage is probably most important at the foot of higher and better drained river-banks and dunes, traditionally used for human settlement and crops. Irregular high stream discharges, combined with high water levels in the River IJssel and supply of water through the Baakse Overflow (De Jong, 1949) induced prolonged high ground water levels and flooding of the flat river-plain until about 1960 (Minderman, 1981). As a result of river improvement, drainage and reallotment schemes, inconveniences due to excess water have ceased since 1960. Now, flooding seldom occurs and then only for short periods.

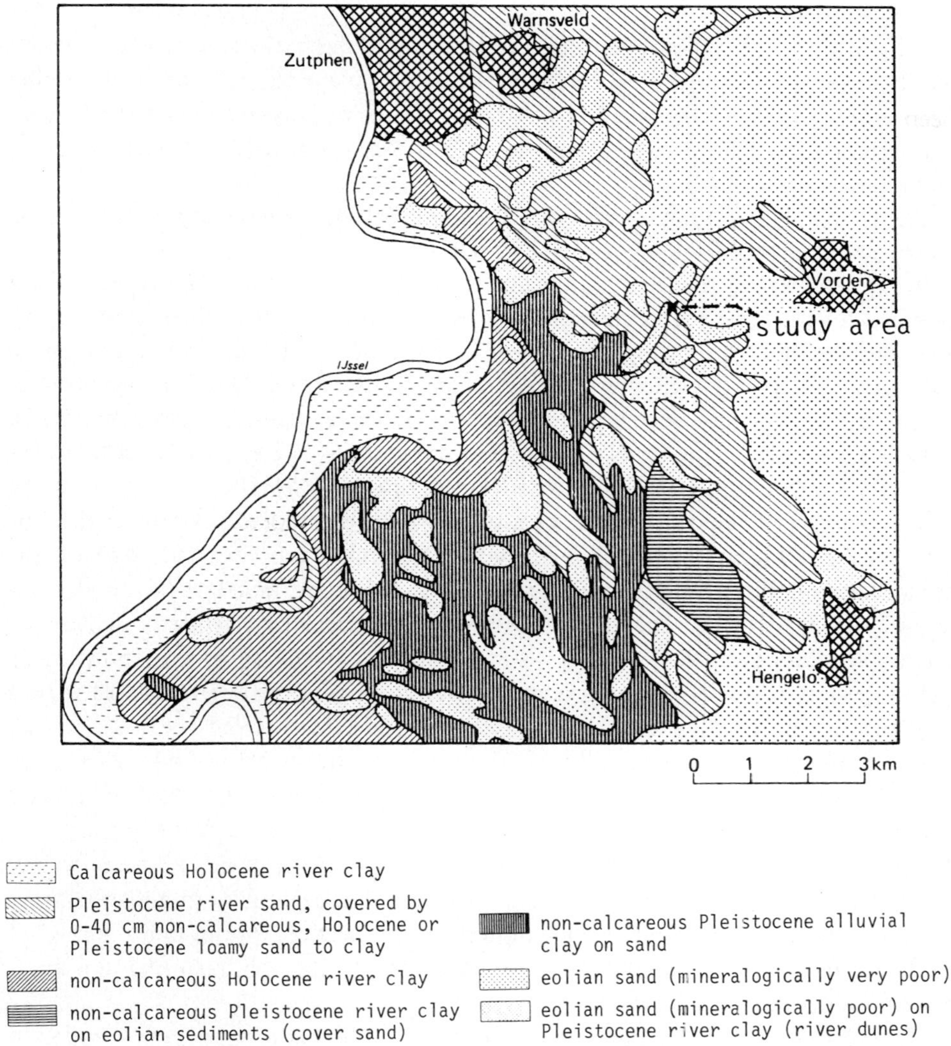

Figure 3. The distribution of eolian and alluvial sediments east of the river IJssel. From: Bodemkaart van Nederland 1:50.000, 33 W/O Apeldoorn (Stiboka, 1979).

2 Soils

2.1 Soil development in the region *(A.G. Jongmans & W.J.F. Visser)*

With low sea-levels during the Pleistocene water tables in the Netherlands were lower than at present, and most of the braided river plain was well drained. Higher sea-levels in the Holocene combined with excess precipitation over evaporation caused periodically high water tables in the lower areas, while higher adjacent areas remained well drained.

The following hypothetical account of soil development in the region is based on an interpretation of data from various sources.

In the whole area, decalcification due to leaching of calcium carbonate has occurred since the late Pleistocene (van Dis, 1984). Soils were decalcified to the greatest depths in the higher areas (Heymans & Tuinhof, 1983). After decalcification, clay migrated, resulting in clay lamellae in sandy sediments (Distel & van Straten, 1980), and weakly developed argillic B horizons in clayey sediments (van Dis, 1984). Such clay illuviation took place at the end of the Pleistocene (Hoeksema & Edelman, 1960; Miedema et al., 1978; van Vliet & Langohr, 1981).

Since the beginning of the Holocene, soils at the well drained sites have been perforated by soil fauna. Thus, sedimentary stratification was obliterated, allowing the vegetation to develop an intensive root system to the lower boundary of the perforated soil (80 cm depth). Leaching of bases was moderated by efficient nutrient cycling by the intensive root systems. At well drained positions brown forest soils, ('Holt' podzol soils) and in the lowest areas humic gley soils ('Beek' earth soils) were formed.

The rise in ground water during the Holocene caused the area of gley soils to increase. Many of the brown forest soils became periodically influenced by ground water, resulting in accumulation of iron oxides in the profile and in a significant reduction of the rooting depth of the forest vegetation. As a result, nutrient cycling became less efficient, leading to stronger leaching and eventually to podzolization. Decreased decomposition of organic material under wet conditions and removal of the forest floor by man may have contributed to podzolization, resulting in poorly-drained podzols.

The inflow of iron-containing ground water from higher areas nearby resulted in an appreciable accumulation of iron in hydromorphic soils. Presumably with the presence of these large amounts of iron at a shallow depth, the podzol profiles remained shallow. In the lowest parts of the landscape or where seepage from adja-

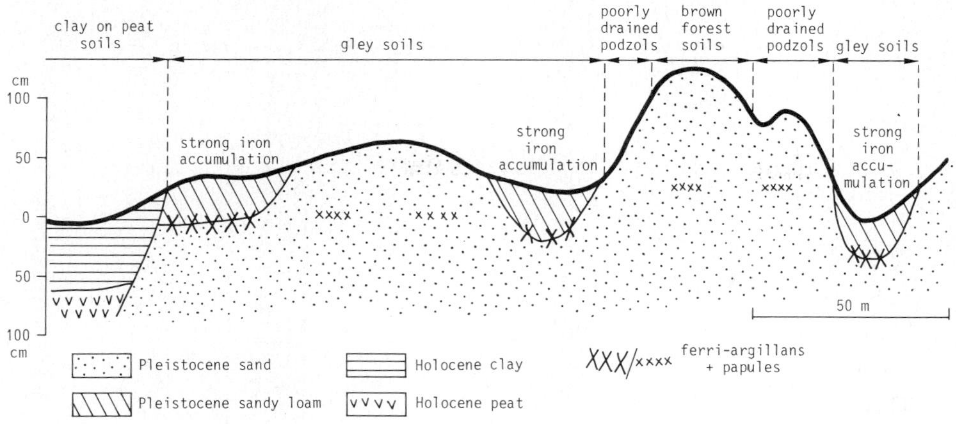

Figure 4. Hypothetical cross section of the landscape, illustrating the relationship between soils and topography.

cent banks or river dunes occurred, the transported iron was concentrated in a thin zone with ironstone (Knibbe, 1969). Figure 4 shows a hypothetical west-to-east cross-section through the landscape with the various soils developed in the sediments.

Man has interfered in many ways in the soil-forming processes. By constructing raised beds and ditches ('rabatten' system) for coppicing, calcareous subsoil material has been brought to the soil surface locally (Jongmans, 1980; 1981). Calcareous surface soils may also have formed as a result of mining ironstone as an ore or as material for constructing churches and city ramparts (Distel & van Straaten, 1980; van Breemen & Jordens, 1983).

Starting from the end of the 19th Century, most of the river plain was cultivated. To improve the productivity of the land, sand was removed from higher areas to lower sites in order to fill depressions, both on a large and on a small scale. In addition, watertables were lowered by drainage, especially from 1960 onwards (Minderman, 1981).

2.2 Soil development in the 'Oude Maat' *(A.G. Jongmans & W.J.F. Visser)*

Within the study area, four soil types have been distinguished and were used to select the measuring plot: A, B, C and D. The toposequential relationships among the plots are shown in Figure 6. The situation of the plots and the soils are indicated in Figure 7. Profile descriptions of soils representative for these plots are given in Appendix 1. Some general morphological and chemical characteristics of the soils are summarized in Figure 5. Detailed descriptions of the soils and their chemical and physical properties are given in Sections 2.3-2.5. Here we only discuss some general soil characteristics and give a hypothetical account of the soil development in the

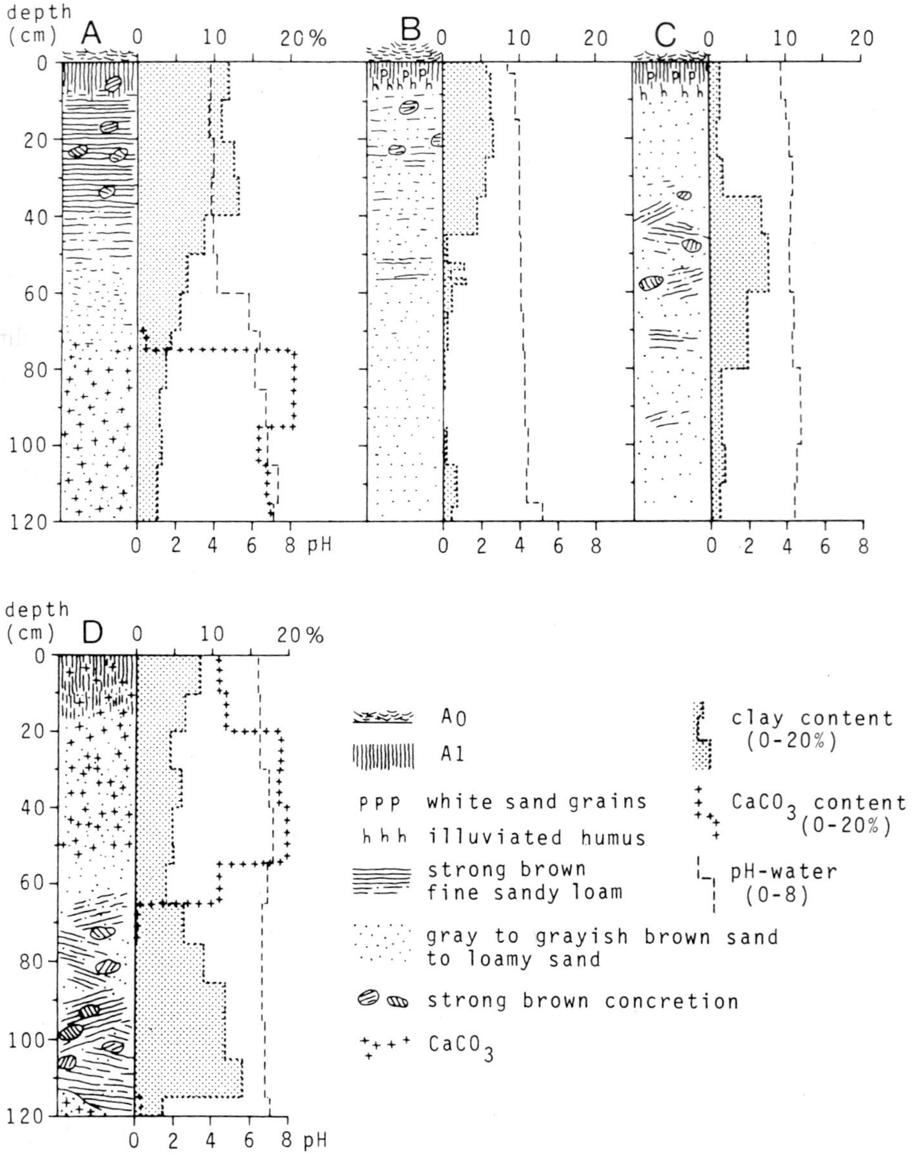

Figure 5. Some morphological and chemical characteristics of the soils used in this study (From: Van Breemen and Jordens, 1983).

Table 1. Classification of the soils of the monitored plots: nearest equivalents at the family level of Soil Taxonomy (USDA, 1975).

Plot A	coarse loamy, mixed, acid, mesic Aeric Haplaquept
Plot B	sandy, mixed, acid, mesic Umbric Dystrochrept
Plot C	mixed, acid, mesic Aquic Udipsamment
Plot D	sandy, mixed, calcareous, mesic Typic Haplaquoll

'Oude Maat'. Table 1 gives the classification of the soils according to Soil Taxonomy (USDA, 1975).

Studies by Distel & Van Straten (1980), Buddingh & Vissers (1981) and Heymans & Tuinhof (1983) clearly show that in the area surrounding the 'Oude Maat', the original soil and land surface has been greatly altered by man's activities, including digging, transport of earth, and formation of plaggen epipedons. Only in a limited part of the area the original topographical situation is still intact.

To understand the genesis of the four soil types and their relation to the topography, the original situation must be reconstructed as far as possible. The following description is a hypothetical reconstruction of the original soil conditions of the 'Oude Maat' and the alterations brought about by man. The reconstruction is based on extensive field evidence and the results of a very detailed soil survey by Waenink (reported by Minderman, 1981).

In the original situation, Plots A and D were the lowest (Fig. 6). Both soils had a

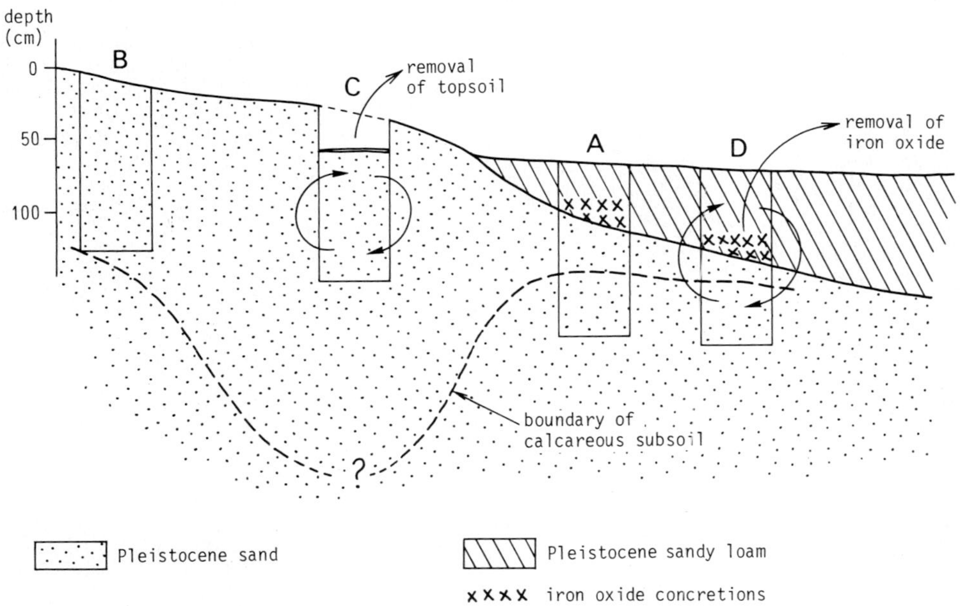

Figure 6. Hypothetical reconstruction of the original soil conditions at the research plots A, B, C and D, and the human activities that affected the soil profiles.

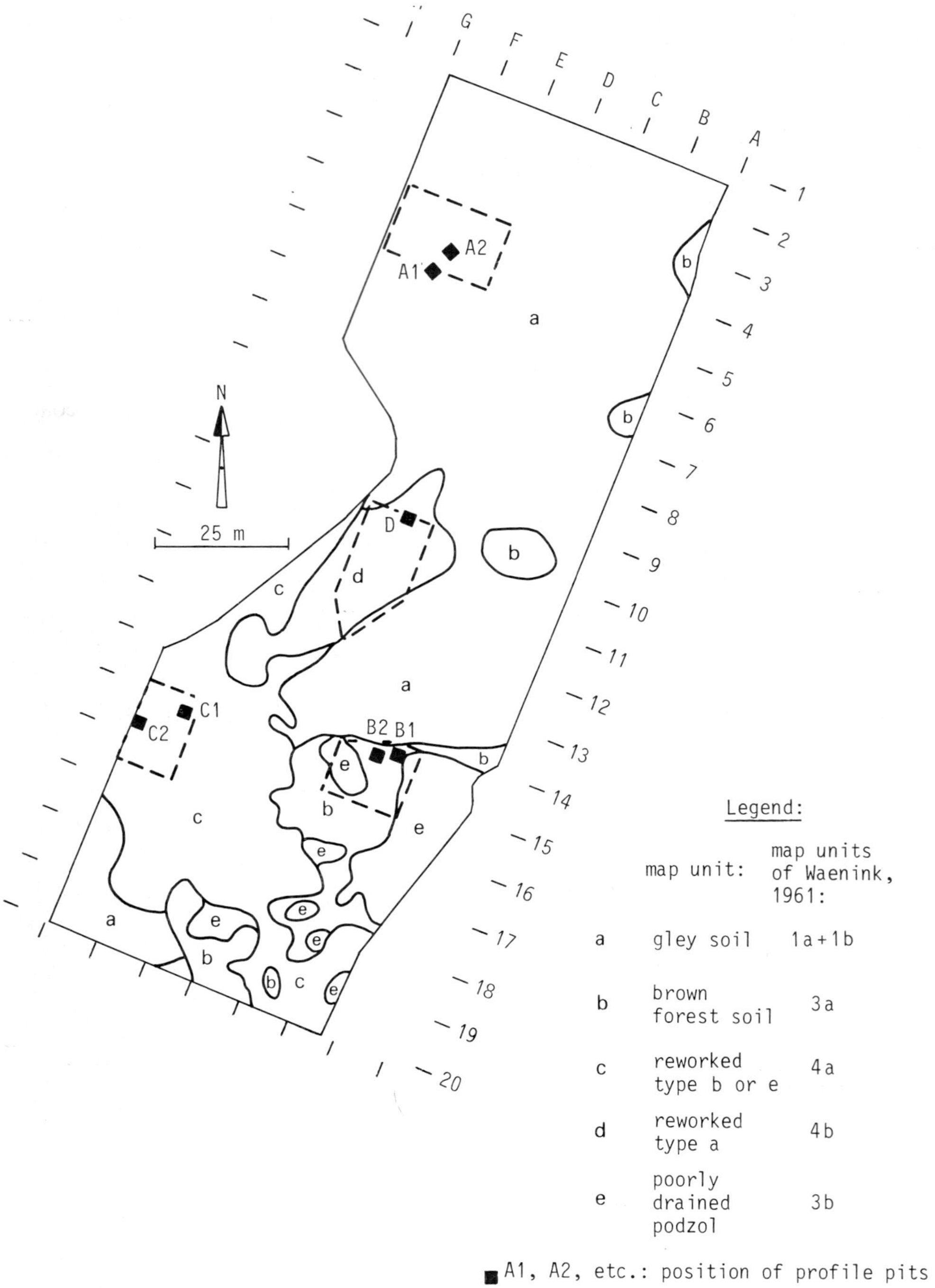

Figure 7. Distribution of soils in the 'Oude Maat' based on the soil map by Waenink (in Minderman, 1981).

calcareous subsoil at shallow depth (70-100 cm) and a strong accumulation of ironstone. In Plot D, this ironstone was mined and the profile was turned upside down in the process. As a result, the calcareous subsoil is now at the soil surface. The undisturbed version of Soil D can be seen in Plot A, although Soil D may have contained more ironstone than Soil A.

Plot B has not been disturbed by human activities except perhaps shallow digging or removal of the litter layer. It occupies about the highest position in the woodland and lies about 25 cm above Plots A and D. It was decalcified to a depth of 130 cm and has some iron oxide enrichment in the subsoil. This 'Holt' podzol soil (brown forest soil) has some clay illuviation at 40-60 cm depth, and a microhumus podzol in the topsoil (Van Dis, 1984).

Profile C resembles Profile B in many respects, but the top 30 to 40 cm have been turned upside down. Depth of decalcification is at least 250 cm. In the present coarse sandy topsoil (the former subsoil), a microhumus podzol has developed. Because its present height is very similar to that of Plots A and D, we assume that about 20 cm of soil material has been removed, perhaps when the profile was turned upside down. We do not know when these digging activities took place. However, because all soils have distinct A horizons and rather regular organic carbon profiles, we assume that several centuries may have elapsed since the disturbance.

The reconstructed soil units and their original topographical position appear to correspond with the general soil-landscape relationships discussed in Section 2.2.1. However the poorly drained podzols or 'veld' podzol soils are missing. Perhaps a shallowly developed poorly drained podzol has been present in Plot C and has been removed from the soil by men, after which coarse-grained subsoil was raised to the surface. The practice of spreading subsoil material over the surface was common in the coppice-system to minimize growth of weeds (E.C.J. Ott, 1984, Agricultural Univ. Wageningen, pers. comm.). Evidence for a former podzol in Plot C is the great depth of decalcification: Buddingh & Vissers (1981) found decalcification to relatively great depths under poorly drained podzols. Moreover, Distel and Van Straten (1980) found poorly drained podzols in land next to Plot C.

The distribution of soils in the 'Oude Maat' is illustrated in the soil map (Fig. 7), which is based on the very detailed survey by Waenink (in: Minderman, 1981), who made 25 borings in each 100 m^2 in the southwestern part of the surveyed area, and 1 boring in each 100 m^2 in the northeastern part. The mapping units of Waenink have been translated into the soil types discussed in Sections 2.1 and 2.2, as shown in the legend of Figure 7.

2.3 Soil macro- and micromorphology *(A.G. Jongmans & W.J.F. Visser)*

In each plot, one (D) or two (A, B, C) pits were dug. Disturbed and undisturbed samples were taken for analysis and profile walls were described in detail. In the following subsections, techniques for the morphological, chemical, mineralogical and physical studies will be described briefly. The data will be given in appendices and some conclusions will be summarized briefly.

In each pit, two profiles were described according to FAO (1977). Large (8 cm × 15 cm × 5 cm) undisturbed samples were taken for thin sections. Profile descriptions are given in Appendix A. Micromorphological descriptions are given by Van Dis (1984); her results are summarized here in diagrammatic form (Appendix B).

Soil A is characterized by a strong brown loamy surface soil (0-70 cm) over gray calcareous sand (> 80 cm) with a somewhat mottled zone between. The original sedimentary stratification is practically undisturbed below 75 cm depth. Strongly developed features of soil formation are browning of mineral grains at all depths above the calcareous subsoil, illuviated clay at 20-40 cm depth as ferri-argillans, and accumulation of iron in ferric nodules (diameter up to 5 mm) particularly in the clayey part of the soil profile. Ferric nodules seem to have formed over the illuviated clay. The calcareous subsoil contains secondary calcium carbonate in the upper part, and increasing amounts of glauconite grains and iron oxide framboids (suggesting a pyrite origin) with depth.

The mineral surface soil has a very loose structure and is covered with a forest floor, which is 3 to 5 cm thick. Roots are confined mainly to the upper 50 cm with very few vertical roots penetrating to a greater depth.

Soil B is higher, sandier and more deeply developed than Soil A. The matrix is more yellowish (brown) and the amounts of illuviated clay and ferric nodules are much smaller than in Soil A. In the surface soil, some disperse humus and the occasional presence of bleached sand grains indicate incipient podzolization. The subsoil below 100 cm is complex, with varying textures and disturbed layering, indicative of former cryoturbation. Below 130 cm, the subsoil is generally calcareous. Although, as in Soil A, roots are mainly confined to the surface 50 cm, more roots than in Soil A penetrate to a greater depth.

In *Plot C,* reworked and presumably largely undisturbed soils can be found adjacent, with the reworked soil dominant. The undisturbed soil is similar to that found in Plot B. The reworked soil is characterized by a layer 20 to 40 cm thick homogeneous non-loamy pale brown medium sand (apparently subsoil material) on top of yellowish brown finer sand with few ferric nodules and some illuviated clay. Glauconite grains in the surface soil strongly indicate that this material originated from the subsoil. Roots are confined to the surface 50 cm in the undisturbed soil, but are found to greater depth (locally to 100 cm) in reworked soil. Some incipient podzolization is apparent in the surface soil.

Soil D is characterized by a cover of greyish calcareous subsoil material, that varies in thickness from 20 to 60 cm. The calcareous layer lies on top of strong brown loamier material similar to the 0-10 cm surface soil of A. Boundaries between the various types of material are often abrupt and irregular, reminiscent of spade marks. Glauconite grains and oxidized pyritic iron betray the subsoil origin of the present surface soil. Lower contents of calcium carbonate in the 0-15 cm layer and secondary calcite at somewhat greater depth indicate that appreciable decalcification has taken place in the surface soil.

2.4 Soil chemistry and mineralogy *(N. van Breemen)*

Methods

Samples of the mineral soil were taken from pits after profile description in May and June 1980. Several kilograms of soil were taken per horizon or layer from one pit wall over a lateral distance of 0.5-1 m. Samples of the organic forest floor covering 10 cm × 100 cm were taken from 70 sites surrounding Plots A, B and C (Winkels, 1985). Air-dried mineral soil was crushed and sieved (< 2 mm). The fraction of particle size > 2 mm was weighed and classified as (mainly) ferric oxide nodules or (mainly) gravel. All further analyses of the mineral soil refer to the fine earth (< 2 mm) fraction or the clay fraction separated from the fine earth. The pH was determined in an extract with 2.5 litres of water or 1 KCl solution (1 mol/lb) from each kilogram of air-dried soil. Most of the methods are described by Begheijn (1980). Organic carbon in the mineral soil was measured potentiometrically as CO_2 after wet combustion in a mixture of phosphoric acid and chromic acid. Soil nitrogen was transformed with sulphuric acid (into ammonia) and selenium, was steam distilled into boric acid and back-titrated ammonium phosphate. Samples of forest floor were ovendried for 36 h at 105 °C; organic carbon was estimated from loss on ignition (1.5 h at 850 °C) and organic N as NH_4 with salicylate blue after destruction in a mixture of sulphuric acid and selenium salicylic acid (Houba et al, 1979). 'Free' Al_2O_3 and Fe_2O_3 refer to extractable fractions with ditionite and oxalate, assayed spectrometrically with O-phenanthroline (Fe) and pyrocathechol violet (Al). Amorphous Fe_2O_3 was extracted with oxalic acid and ammonium oxalate in the dark and measured by atomic absorption spectrometry. Carbonate ($CaCO_3$) was estimated potentiometrically as CO_2 evolved after adding a solution of $FeCl_2$ and HCl.

The cation-exchange capacity (CEC) and exchangeable cations were measured by Begheijns' (1980) method with LiEDTA for calcareous soils, and Bascomb's (1964) with $BaCl_2$ and $MgSO_4$ (unbufferd) method for non-calcareous soils. Elemental analysis was by X-ray fluorescence spectroscopy of a lithium-tetraborate melt of the soil material, except for FeO and MgO, which were analysed colorimetrically with O-phenanthroline (Fe) or by atomic absorption spectroscopy (Mg) after destruction with HF and H_2SO_4 for 10s.

Clay was separated after removal of carbonates and of organic matter with resp. Na-acetate-acetic acid and hydrogen peroxide, and flocculated with Ba-acetate. Elemental analyses refers to the Ba-saturated clay fraction. Assuming a cation exchange capacity of 500-1000 mmol per kg of clay, the BaO content (mass fraction) of Ba-saturated clay should be 3.8 to 7.6 %. Occasional higher values indicate incomplete removal of Ba-acetate before analysis. Particle size analysis was done at the 'Laboratorium voor grond- en gewasanalyse Mariëndaal', Oosterbeek, by sieving and gravity-sedimentation after treatment with hydrochloric acid and hydrogenperoxide to remove carbonates and organic matter.

X-ray diffraction of oriented clay separates on porous ceramic tiles refer to Mg-saturated clay with and without glycerol and to K-saturated clay after drying and

Table 2. Soil chemical data. Mass fractions refer to oven-dried soil.

	depth	mass fraction (%) C	N	free Fe_2O_3	$CaCO_3$	day	pH-H_2	pH-KCl	CEC (mmol/kg)	base-saturation (%)
A2	5-0	42	1.9	–	–	–	–		–	–
	0-8	3.1	0.20	5.2	0.0	11	3.7	3.4	50	3
	8-20	0.9	0.07	6.3	0.0	11	3.6	3.4	46	1
	20-30	0.4	0.03	11.6	0.0	14	3.7	3.6	–	–
	30-40	0.3	0.02	1.1	0.0	15	4.2	3.7	42	28
	40-50	0.1	0.01	1.7	0.0	7	4.7	3.8	31	63
	50-60	0.1	0.01	0.7	0.0	5	4.5	4.0	29	67
	65-80	0.2	0.01	0.9	0.1	5	6.1	5.9	33	91
	80-100	0.1	0.02	0.3	17.4	2	7.1	8.1	20	100
B2	5-0	47	2.1	–	–	–	–		–	–
	0-4	6.6	0.18	2.2	0.0	7	3.8	3.0	145	20
	4-8	3.0	0.07	2.4	0.0	8	3.7	3.2	62	8
	8-20	2.5	0.08	2.3	0.0	7	4.0	4.0	31	3
	20-30	1.5	0.03	2.5	0.0	7	4.1	4.2	20	2
	30-40	0.5	0.02	2.9	0.0	6	4.2	4.3	15	3
	40-48	0.3	0.01	2.7	0.0	7	4.2	4.2	16	1
	48-60	0.2	0.01	1.0	0.0	4	4.3	4.2	11	2
	60-100	0.1	0.01	0.4	0.0	1	4.5	4.3	7	5
	100-120	0.2	0.01	0.9	0.1	6	6.8	6.1	39	100
C2	4-0	41	1.8	–	–	–	–		–	–
	0-10	2.3	0.14	0.7	0.0	2	3.8	3.3	30	5
	10-15	0.3	0.02	0.5	0.0	2	4.1	3.8	15	3
	15-25	0.2	0.02	0.7	0.0	1	4.2	3.8	14	2
	25-35	0.2	0.01	0.6	0.0	2	4.4	3.9	13	2
	35-45	0.4	0.02	4.6	0.0	7	4.3	4.0	19	< 1
	45-60	0.5	0.02	4.6	0.0	7	4.2	4.1	18	< 1
	60-70	0.1	0.01	1.4	0.0	5	4.4	4.0	15	4
	70-80	0.1	0.01	2.4	0.0	5	4.3	3.9	17	13
	80-90	0.1	0.01	0.8	0.0	1	4.7	4.0	13	4
	90-100	0.1	0.01	0.6	0.0	2	4.7	4.1	13	16
D1	0-10	3.3	0.27	0.6	10.9	8	6.4	7.2	64	100
	10-20	3.0	0.20	0.5	11.4	7	6.5	7.4	49	100
	20-30	0.4	0.04	0.5	19.1	5	6.5	7.8	20	100
	30-40	0.2	0.01	0.5	18.9	6	7.0	8.1	14	100
	40-50	0.1	0.01	0.5	19.8	5	7.2	8.2	17	100
	50-55	0.1	0.0	0.6	11.0	5	7.2	8.1	16	100
	55-65	0.2	0.01	1.1	0.3	4	6.9	7.7	25	100
	65-75	0.3	0.01	2.4	0.0	6	6.6	7.0	25	100
	75-85	0.1	0.02	7.1	0.1	9	6.6	7.1	37	100
	85-95	0.1	0.03	1.1	0.2	12	6.7	7.2	44	100
	95-105	0.1	0.04	1.5	0.5	12	6.8	7.7	62	100
	135-145	0.1	0.01	0.2	19.2	5	7.4	8.3	13	100

heating to various temperatures. 1.4 nm clay was classified as smectite (expansion to 1.8 nm upon glycerol treatment of Mg-clay), as hydroxy-interlayered vermiculite (incomplete collapse to 1.0 nm after K-saturation, collapse to 1.0 nm after heating to 300-550 °C), as vermiculite (collapse to 1.0 nm after K-saturation), as chlorite (no change in the 1.4 nm peak after K-saturation or heating to 550 °C) or as interstratified chlorite/vermiculite (peak between 1.0 and 1.4 nm after heating to 300-550 °C of a K-saturated sample). Some information on the mineralogy of the sand fraction has been derived from micromorphological observations.

Results

Detailed soil chemical data can be found in the microfiche of which some results have been summarized in Table 2. The most important features will be discussed briefly below. Undisturbed soil profiles and their disturbed 'counterparts' will be treated in that order.

Profile A

The two soil profiles sampled and analysed here are very similar. The forest floor is about 5 cm thick and of the mor type, with a C/N ratio of 22.3 (Table 3). The mineral soil has fine sandy loam texture in the surface 40 cm, grading through loamy fine sand between 40 and 50 to 60 cm depth to fine sand below 50 to 60 cm depth.

Organic matter decreases regularly with depth (Fig. 8) from 3% at 1-10 cm depth to 0.1-0.2% below 40 cm depth.

The surface soil is strongly acidic (pH-H_2O between 3.5 and 4). Between 40 and 80 cm the pH gradually increases to near-neutral values in the calcareous subsoil (15-20% $CaCO_3$) which starts abruptly at 75 to 80 cm below the surface.

The cation exchange capacity is low (40-100 mmol. kg^{-1}) and is mainly caused by organic matter (mean CEC 2.1 mmol. g^{-1} of org C, for all acidic soil samples from the plots A, B and C). Except in the and just above the calcareous subsoil, Al^{3+} and H^+ dominate the adsorption complex, with slightly more bases near the surface than at 10-40 cm depth.

Total sulfur contents vary from 0.18% in the forest floor to less than 0.01% in the subsoil (Table 4). At shallow depth sulfur is mainly organic, partly in a very stable form (not extractable by Raney Nickel), whereas at greater depth it is almost all in sulfate-form. The sulfate is probably associated mainly with fine-grained iron oxi-

Table 3. The forest floor: thickness, pools of C and N, and C/N ratio. Values are means ± s.d. based on 70 samples (Winkels, 1985).

plot:	A	B	C
thickness (cm)	4.9 ± 1.0	5.5 ± 1.2	4.4 ± 0.7
C $t.ha^{-1}$	22.3 ± 5.2	25.8 ± 7.2	19.8 ± 3.0
N $t.ha^{-1}$	1.00 ± 0.23	1.14 ± 0.31	0.88 ± 0.13
C/N	22.3	22.7	22.6

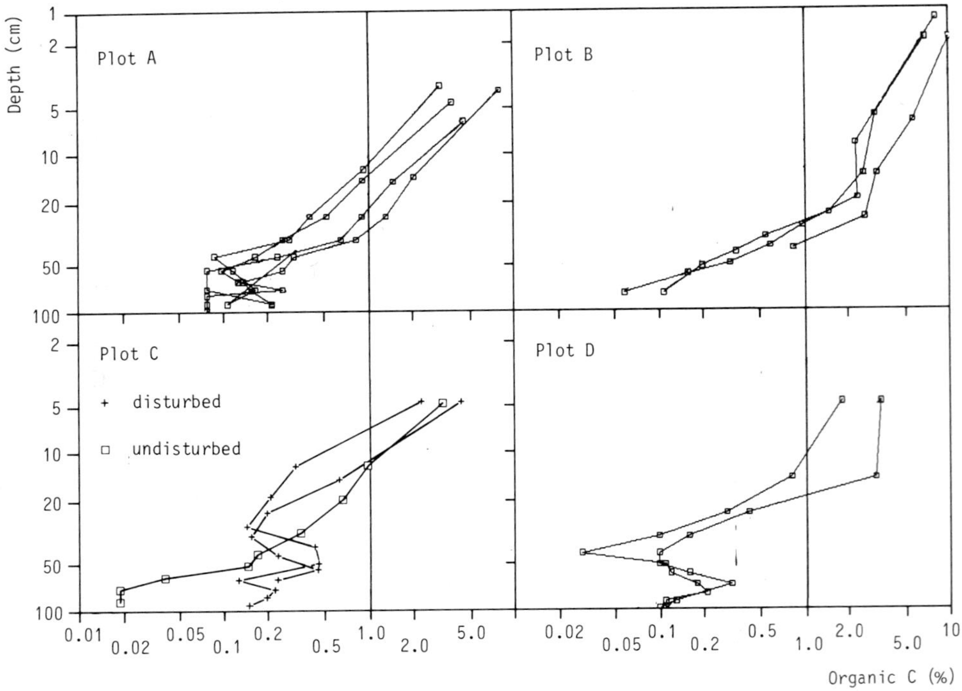

Figure 8. Organic carbon contents of the fine earth against depth for the mineral soil in plots A, B, C and D (double logarithmic scale).

Table 4. Various sulfur forms in soil samples taken from plots A, B and D (S, in g/kg soil). HI-red.: HI-reducible sulfur, mainly inorganic plus organic (SO_4-ester) sulfate. NI-Al: Raney Nickel reducible sulfur, mainly carbon-bonded S. Total: determined by methylene blue after extraction with $NaHCO_3$-Ag_2O. Samples taken in October 1985 and freeze-dried immediately after sampling (Altena, 1986).

plot	A			B			D		
depth (cm)	HI-red.	Ni-Al	total	HI-red.	Ni-Al	total	HI-red.	Ni-Al	total
forest floor	0.27	0.28	1.78	0.33	0.21	2.02	–	–	–
0-5	0.21	0.11	0.84	0.23	0.13	0.92	0.14	0.05	0.32
5-10	0.19	0.08	0.40	0.37	0.14	1.01	0.15	0.05	0.36
10-20	0.23	0.04	0.24	0.19	0.08	0.48	0.10	0.04	0.24
20-40	0.33	0.03	0.31	0.21	0.05	0.23	0.09	0.04	0.46
40-60	0.25	0.03	0.17	0.12	0.03	0.15	0.02	0.02	0.06
60-100	0.11	0.03	0.09	0.04	0.03	0.30	0.03	0.02	0.07
100-130	0.04	–	0.11	0.05	0.03	0.24	–	0.00	0.08

Table 5. Clay mineralogy of selected samples from soils of the four plots. 1.4 nm minerals are specified as follows: C = Chlorite, V = Vermiculite, C/V = Chlorite-Vermiculite interstratified, S = Smectite. +-+ + + + + = little-abundant.

profile	depth (cm)	Abundance based on the intensities of (001) reflections			
		0.7 nm	1.0 nm	1.4 nm	
A1	0-10	+	–	+ + + +	C/V
	10-20	+	–	+ + + +	C/V
	20-30	+	–	+ + + +	C/V
	30-40	+	(+)	+ + + +	C/V, V
	40-50	+	+	+ + + +	V, C/V
	50-60	+	+ +	+ + + +	V, C/V, C
	60-70	+	+ + +	+ + +	V, C/V, C
	85-125	+	+ + +	+ + +	C, V, S
B2	0-4	+	–	+ + + + +	S
	4-20	+	–	+ + + + +	V, C/V
	30-48	+ +	+	+ + + +	C/V
	48-100	+	+ +	+ + +	V
	120-140	+	+ +	+ + +	S
C1	0-15	+(+)	+	+ + + +	V, C/V
	15-40	+(+)	–	+ + + +	C/V
	40-70	+ +	+(+)	+ + +	C/V, C
D1	0-10	+	+ + +	+ +	C, V
	30-40	+	+ + +	+ +	C, V, S
	85-95	+	–	+ + + +	V
	105-115	+	–	+ + + +	V
	135-145	+	+ + +	+ + +	C, V, S

des. Altena (1985) found high correlations between HI-reducible sulfur and oxalate extractable ('amorphous') iron: r = 0.99 at A and 0.90 at B.

The clay mineralogy shows a pronounced change with depth (Table 5). Especially in the upper half of the profile, 1.4 nm minerals are predominant whereas illite (1.0 nm) is absent in the surface soil and increases with depth below 40 cm. The 1.4 nm minerals are mainly randomly interstratified chlorite/vermiculite at shallow depth, chlorite plus vermiculite at greater depth, with some smectite in the calcareous substratum.

Variation in clay mineral composition with depth may be due largely to variations in sediment composition, rather than to weathering and clay transformation. The abrupt change in the chemical composition of the sand + silt fraction, and the clay content at 40 cm depth (Fig. 9) must be ascribed to the presence of different sediments above and below 40 cm depth. The uniformity in chemical composition of sand + silt in the top layer suggests that silicate minerals in this fraction have been little influenced by chemical weathering. The clearest sign of weathering is the abrupt appearance of $CaCO_3$ at 75-80 cm depth in an otherwise homogeneous sediment: there can be little doubt that the 40-80 cm layer has been decalcified in the course of soil development.

Other aspects of soil development, such as clay illuviation (mainly at 30 to 40 cm depth, see Fig. 9) and accumulation of ferric iron in the surface horizons have been discussed. Fig. 10 illustrates the positive correlation between the contents of clay and of free ferric oxide in soil samples with more than 4% clay. Similarly, the content of iron oxide concretions coarser than fine earth are positively correlated to the clay content. Presumably, clay has favoured the accumulation of iron by adsorption of Fe^{2+} from upwelling groundwater.

Profile D

As discussed earlier, profile D is an 'upside down' version of profile A. This is clearly illustrated by the chemical data. Where the calcareous surface soil is quite thick, as in D_1 (south) the soil at 20 to 50 cm depth has practically the same $CaCO_3$ content as the subsoil of plot A, and of the undisturbed deepest parts of profile D, viz. 16-26% $CaCO_3$. The present 75-115 cm layer of D_1 south (missing in D_1 north) is very similar to the A_1 surface soil in practically all respects, except that the pH is near-neutral and the adsorption complex is saturated with Ca^{2+}, both of which must

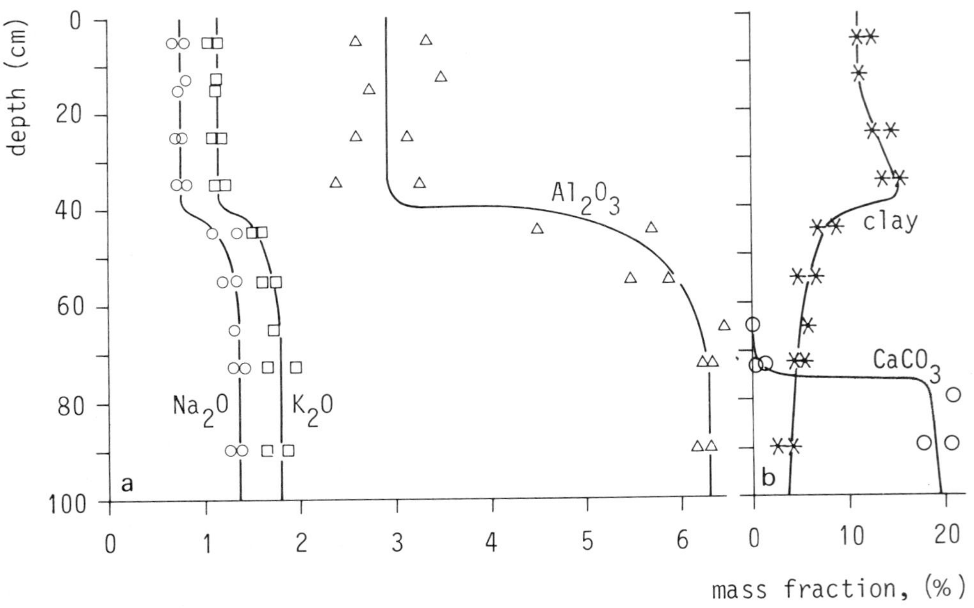

Figure 9. Distribution with depth in soil profiles A1 and A2 of:
a. contents of Na_2O, K_2O and Al_2O_3 in the sand + silt fraction (free of $CaCO_3$ and organic carbon, calculated from elemental composition of fine earth and clay, and of the contents of clay, organic C and $CaCO_3$ in the fine earth), and
b. contents of clay (< 2 μm) and $CaCO_3$ in the fine earth.

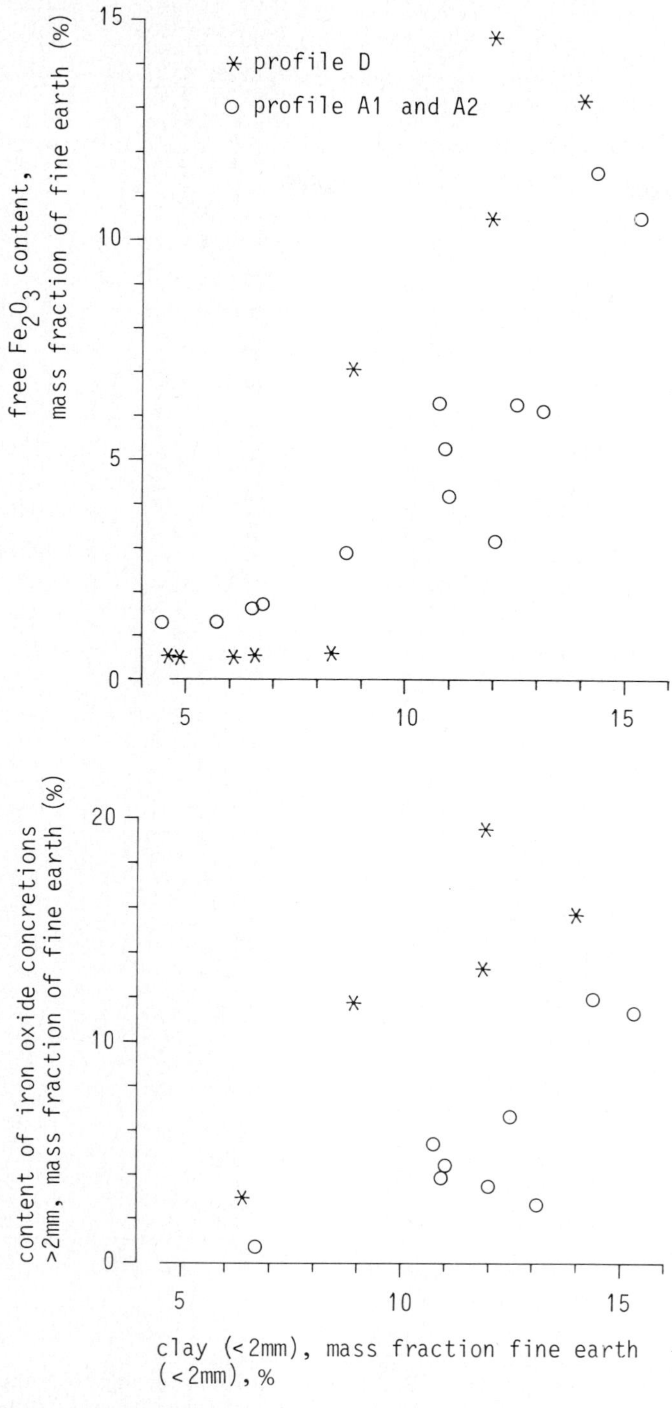

Figure 10. Relation between clay content and the contents of free Fe_2O_3 and iron oxide concretions in soils A and D.

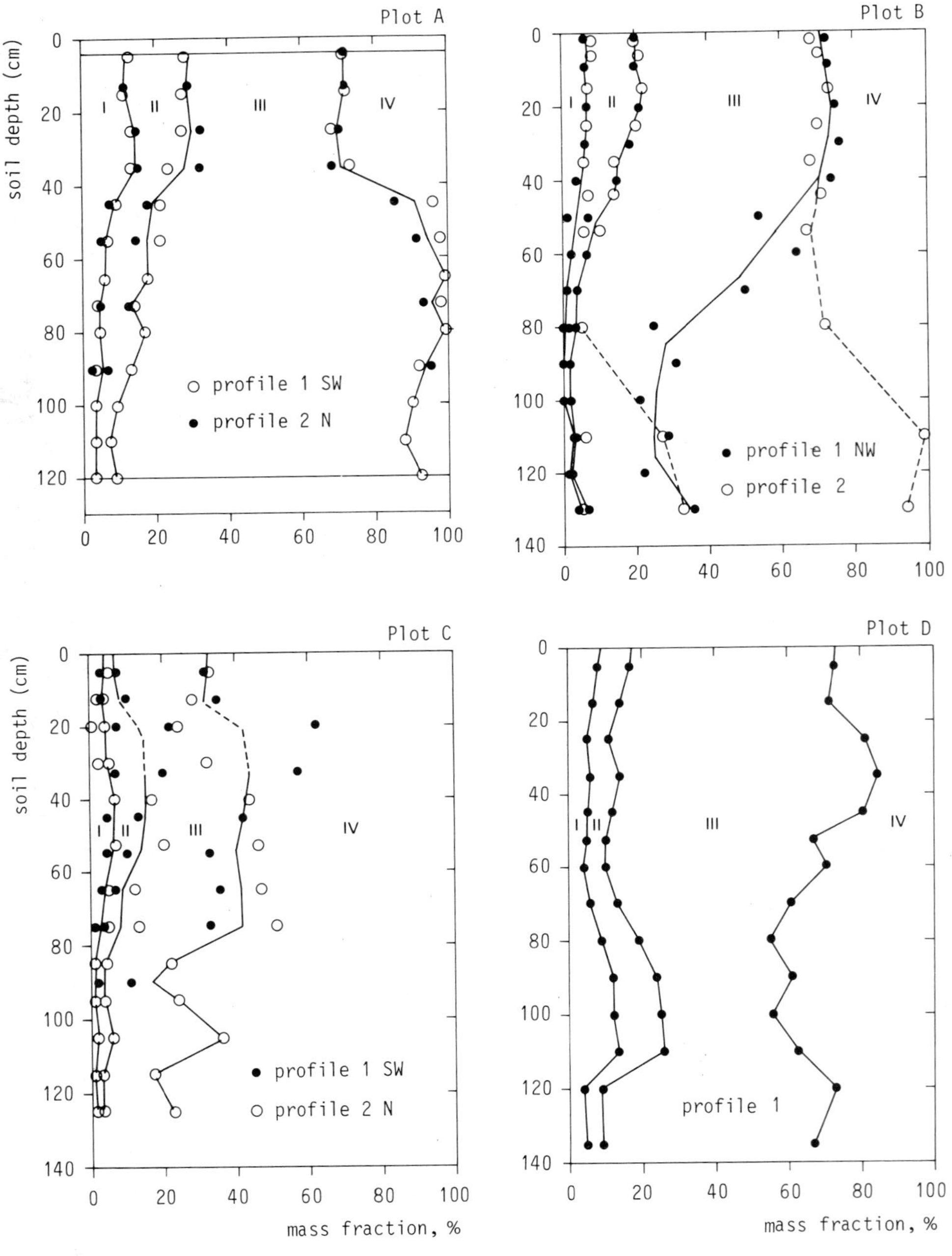

Figure 11. Textural profiles for plots A, B, C and D. Sides of the profile pits sampled are indicated as SW (southwest), N (north) and NW (northwest). Textural ranges distinguished are clay (< 2 μm, I), loam (2 – 50 μm, II), fine sand (50 – 210 μm, III), and coarse sand (210 – 2000 μm, IV).

be attributed to the overburden of calcareous soil material. The same layer, however, shows a distinctly higher ratio of Fe_2O_3 to clay than the surface soil of A (see Fig. 10). This corroborates the hypothesis that profile D was used for mining of iron. The presence of a well-developed Al horizon (without forest floor) indicates that sufficient time has passed since the disturbance took place to build up a more or less natural organic carbon profile. However, C-contents vary irregularly with depth, which may reflect earlier digging activities (see Fig. 8). Total sulfur contents are distinctly lower than at A and B, and no significant correlation was observed between HI-reducible (sulfate) S and iron oxides (Altena, 1985).

Profile B

At plot B, the mor-type forest floor is somewhat thicker than at A (Table 3). The textural profile is 40 cm of loamy sand (with 6-8% clay) over sand (median mainly 105-210 μm), with layers or pockets of fine and very fine sand or very fine sandy loam below 60-80 cm depth. The soil pH (in H_2O) is close to 3.5 in the first 10 to 20 cm and increases gradually from 4 to 5 between 20 and 100 cm depth, reaching near-neutral values in loamy, calcareous pockets in the subsoil. Organic carbon profiles in the two soil profiles studied are virtually identical. An apparent accumulation of organic carbon between 10 and 20 cm depth (Fig. 8) may indicate the beginnig of a humus podzol B horizon. The presence of smectite instead of chlorite/vermiculite in the 0-4 cm surface horizon also points to podzolization. Otherwise, the clay mineralogy is similar to that of profile A: a dominance of interstratified chlorite/vermiculite at shallow depth, of vermiculite at greater depth and of smectite in the calcareous subsoil. Appreciable amounts of illite occur only below 50 cm depth. As in A, the contents of Ca, Mg, K and Na in the sand + silt fraction tend to increase with depth more or less parallel with changes in texture. Therefore the low contents of base cations at shallow depth can not be taken mainly as signs of weathering of the surface horizons.

Profile C

The forest floor is thinner than at plots A and B, and contains distinctly less carbon and nitrogen (Table 3). Loamy and finer textures are lacking and the median of the sand fraction is between 210 and 420 μm (Fig. 11). Irregular organic C-profiles with anomalously high contents at 35 to 60 cm depth (0.4-0.5%), may reflect the disturbance of the soil profile (Fig. 8). Although the texture is coarser than of soils A and B, the total base content (or acid neutralizing capacity) of the upper 40 to 50 cm of the soil profile is appreciably higher, which may help to explain the distinctly higher soil pH (3.5 to 4). The higher pH and higher base status may be partly the result of bringing mineralogically richer (less weathered?) soil material to the surface.

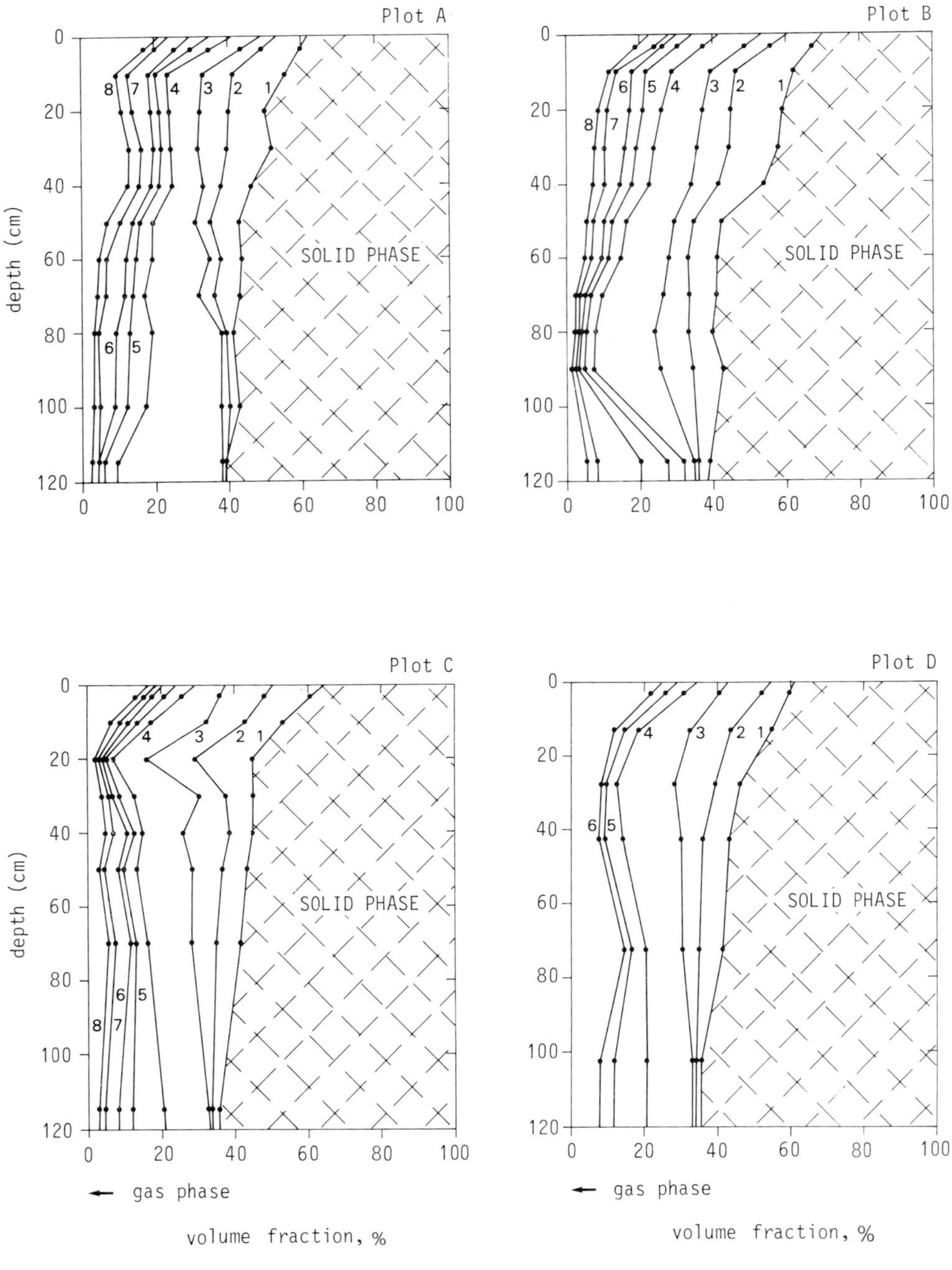

Figure 12. Volumetric water content as a function of pressure potential (mbar) and depth for plot A, B, C and D. Volumetric water contents are plotted for the following pressure potentials:
(1) – 0, (2) – 10, (3) – 32, (4) – 100, (5) – 200, (6) – 501, (7) – 2500 and (8) – 16000 mbar.

2.5 Soil physical properties *(J.J.M. van Grinsven)*

Texture

Textures of the Hackfort soils range from sand to sandy loam. Variation in texture with depth is shown in Figure 11 for plots A, B, C and D. Clay contents (< 2 μm) range from 1 to 15%, silt contents (2-50 μm) range from 1 to 30%. The grain size of the sand fraction varies strongly. Soil A is a loamy fine sand, while soil C is a loamy coarse sand. The subsoil at plot B is very heterogeneous with both fine sandy and coarse sandy textures. The strongly variable texture in plot C probably reflects digging activities.

Water retention

Water retentivity functions were established by means of equilibration techniques on 5 cm long and 5 cm wide undisturbed soil cores. Generally 3 to 6 cores were sampled at 10 cm depth intervals for all plots, down to a depth of 1.20 m. Water contents for −10, −32 and −100 mbar pressure potential were obtained by equilibrating the cores with saturated 'Blokzijl' sand (fine sand: main fraction between 35 to 75 μm) for −200 and −500 mbar with saturated kaolinite. Water contents at −2500 and −16000 mbar were obtained by equilibrating small disturbed soil samples (1 cm high, 2.7 cm wide) in a pressure membrane apparatus. Prior to equilibration all soil samples were prewetted on saturated Blokzijl sand at zero pressure potential.

The resulting water retention data are given in Appendix 3 and shown in Figure 12. In all soils, the pore volume gradually decreases from 0.6 at the soil surface to 0.4 at 1.20 m depth. Relatively high pore volumes at shallow depth are due to the abundance of biopores. In the surface soils, biopores of varying diameters, relatively high clay and loam contents (Figure 11), and high organic matter contents, cause a slow and gradual decrease of water content with decreasing pressure potential. The high pore volume and high water retention capacity of the surface soil are important in terms of soil water storage. Water storage in the forest floor can be of importance (Schroeder & Buck, 1978), but its water retention behaviour was not studied.

At depths where both loamy and sandy soil materials were found, water retention data were simply averaged even though both materials were not samples according to their relative abundance.

Retention data were generally based on six soil cores, except at D, where only three cores were used. The anomalous water retention behaviour at 20 cm depth in plot C probably reflects former human digging activities.

Hydraulic conductivity

Saturated conductivities were measured on undisturbed soil cores, 10 cm long and 5 cm in diameter (Table 6). Conductivity at 10 mbar suction was measured for the same cores, using the two plate steady state method (Klute, 1972). Measurements

Table 6. Hydraulic conductivity at saturation (K_{sat}) and at 10 mbar suction (K_{10}) (means and standard deviation).

plot	soil depth (m)	number of samples	hydraulic conductivity ($m.d^{-1}$)			
			K^{sat}	σ	K_{10}	σ
A	0.33-0.43	4	1.06	0.71	$5.86\ 10^{-1}$	$5.54\ 10^{-1}$
	0.60-0.70	4	$3.35\ 10^{-1}$	$2.96\ 10^{-1}$	$1.68\ 10^{-1}$	$6.15\ 10^{-1}$
	1.12-1.22	4	4.50	1.16	3.25	1.14
B	0.49-0.60	4	1.13	0.74	$6.81\ 10^{-1}$	$5.96\ 10^{-1}$
	0.60-0.70	2	1.94	1.36	1.25	0.91
	0.66-0.76	2	$1.03\ 10^{1}$	$3.54\ 10^{-1}$	$1.03\ 10^{1}$	$3.54\ 10^{-1}$
	0.85-0.95	2	$1.40\ 10^{1}$	$7.07\ 10^{1}$	$1.40\ 10^{1}$	$7.07\ 10^{-1}$
C	0.45-0.55	4	7.23	2.59	7.02	5.52
	0.64-0.76	4	$1.33\ 10^{1}$	3.97	$1.33\ 10^{1}$	3.97
	0.70-0.80	2	$1.25\ 10^{1}$	$7.07\ 10^{-1}$	$1.25\ 10^{1}$	$7.07\ 10^{-1}$
D	0.25-0.35	2	2.75	0.64	1.27	1.11
	0.45-0.55	6	3.45	1.64	1.26	0.90
	0.70-0.85	4	1.46	0.84	$1.79\ 10^{-2}$	$1.50\ 10^{-2}$
	1.10-1.20	4	$9.43\ 10^{-1}$	$7.34\ 10^{-1}$	$2.04\ 10^{-1}$	$1.53\ 10^{-1}$

for all sampled layers were carried out at least in duplicate. Unsaturated conductivities for very permeable sandy samples were underestimated, due to an appreciable resistance caused by the experimental setup itself. This resistance was later found to be due to the fine tubing used, and was calculated by a correlative comparison of the saturated and unsaturated conductivities, that were obtained independently. Subsequently, all unsaturated conductivities were corrected for the resistance contributed by the two-plate setup; the corrected values are given in Table 6.

The limited data available provide no evidence for impermeable soil layers that could cause temporary waterlogging. Although no conductivity data are available for the surface soil, no doubt infiltration capacities are sufficiently high to transmit water supplied at the highest precipitation intensities observed: the surface soil is highly porous and there is no field evidence for the occurrence of surface runoff.

Short-circuiting during high flow situations due to the presence of continuous macropores was not investigated. Although many large biopores are present, it is unlikely that short-circuiting of flow between the soil surface and the groundwater takes place. The soil appears permeable enough to prevent saturation near the soil surface, which is a precondition for short-circuiting. Groundwater depth sometimes responded quickly to high-precipitation events, but this phenomenon can be explained by fast drainage in the surrounding agricultural area and high permeability of the very sandy subsoil.

In principle, the profiles studied are favourable for capillary recharge from the groundwater because saturated conductivities increase with depth and groundwater tables are generally shallow.

Groundwater depth

Observed weekly to fortnightly groundwater depths from July 1980 to December 1985 are shown in Fig. 13. Groundwater depths roughly vary between 0.3 and 1.4 m depth at plots A, C and D and between 0.5 and 1.7 m at B. Shallow groundwater events in winter and spring are generally short-lived, as a result of artificial drainage of the general area and control of flooding of the IJssel river in winter. The groundwater-time curves for the four plots are practically parallel with the vertical shift equal to the elevation differences, showing the presence of a horizontal groundwater table below the area.

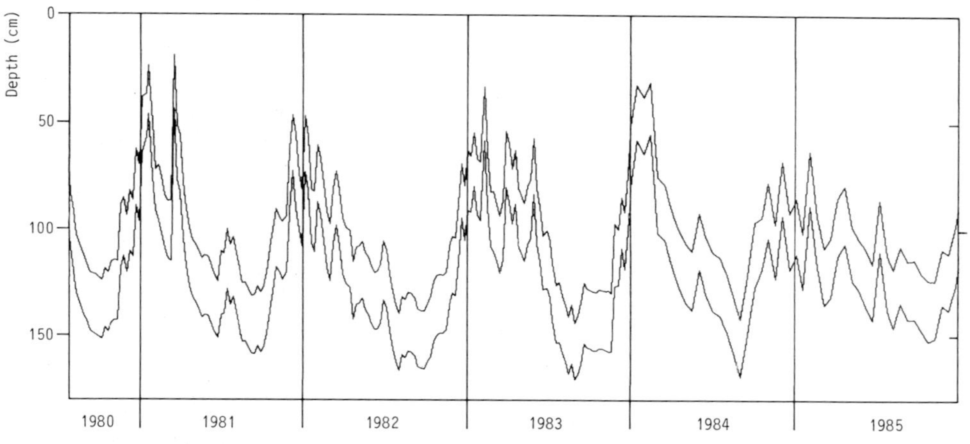

Figure 13. Groundwater depths vs. time. The upper curve shows the mean for plots A, C and D, the lower curve is for plot B.

3 Vegetation

3.1 General description of the vegetation, above-ground biomass and root distribution *(N. van Breemen & P.A.B. de Visser)*

The tree vegetation is dominated by oak (*Quercus robur*) and birch (*Betula pendula*), mainly developed from old coppice-stumps. In the calcareous plot, alder (*Alnus glutinosa*) and poplar (*Populus tremula*) are found as well. The whole area was reportedly coppiced last in 1939, and except for occasional thinning, the vegetation has been left largely undisturbed since then. Table 7 gives some data about the tree biomass at plots A-D.

From all available old maps it can be concluded that the study area has been woodland since at least 1783 (Minderman, 1981). The apparently undisturbed character of the ground vegetation indicates that the 'Oude Maat' indeed has a long history of forest cover. Plot A contains about as much oak as birch, but many birch trees are now (1983-1986) dying, probably as part of the natural succession. Total tree biomass and net annual growth are small, compared to that at B and C (Table 7). The following qualitative description is based on information given by E.C.J. Ott (pers. comm., 1981). Plot A has a sparse but fairly species-rich vegetation with a shrub layer of *Lonicera periclymenum*, *Corylus avellana*, *Frangula alnus* and *Sorbus aucuparia*, and a herb layer of *Pteridium aquilinum*, *Maianthemum bifolium*, *Hedera helix* and *Rubus fruticosus*.

Plot B has a fairly heavy stand of oak, with practically no birch. The ground vegetation is similar to that at A, but denser. *Corylus avellana* and *Hedera helix* are

Table 7. Standing tree biomass, measured in winter 1984/85 and annual growth. Popular and birch were not quantified at D (de Visser, 1986).

	number of stems/ha	basal area m^2/ha	dry mass t/ha	annual growth t/ha.yr
A oak	627	13.9	98	1.7
A birch	366	12.9	45	1.7
B oak	700	27.7	181	5.1
C oak	750	25.0	156	4.4
D oak	650	12.2	63	4.5
D alder	450	6.7	47	1.8

missing, and additional herbs found. These are *Molinia caerulea*, *Anemone nemorosa*, *Stellaria holostea* and *Holcus lanatus.*

Plot C too is an almost pure stand of oak. Although *Corylus avellana*, *Molinia caerulea*, *Rubus fructicosus* and *Hedera helix* are lacking, the vegetation indicates a somewhat 'richer' soil than at A and B, with, in addition to the species observed at A and B, *Prunus serotina, Rubus idaeus, Deschampsia flexuosa,* and *Galeopsis tetrahit.*

Bracken *(Pteridium aquilinum)* is the dominant groundcover during summer at A and B, *Prunus serotina*, *Rubus idaeus*, *Deschampsia flexuosa*, and *Galeopsis te-* *contains Cornus sanguinea*, *Crataegus monogyna*, *Sambucus nigra*, *Ribes nigrum*, *Daphne mezereum* and *Corylus avellana*, while the herb layer includes *Circaea lutetiana*, *Lamium galeobdolon*, *Polygonatum multiflorum*, *Urtica dioica* and *Stachys sylvatica,* plus species found elsewhere, such as *Anemone nemorosa*, *Stellaria holostea* and *Marianthemum bifolium.*

Roots were sampled simultaneously with profile description and sampling of the soil, i.e. in summer 1980. From each plot three or four 25 × 25 cm^2 soil columns were sampled at 5 cm depth intervals, to estimate root biomass and its distribution in the soil profile. The soil was collected in a sharp-edged steel box (25 × 25 × 5 cm^3), open on one side and with several circular holes (to allow easy escape of air and to check the sampling depth) on the other side. The box was pressed with the open side down into the soil and if necessary any horizontal roots were cut using a knife or saw. When the box was pressed down completely, the soil and any roots were cut horizontally just below the box by means of a steel plate. The soil with the roots was removed, and roots were separated by washing and sieving in the laboratory. Living roots (in diameter classes < 1 mm, 1-5 mm, 5-10 mm and > 10 mm) and dead roots were separated as much as possible, air-dried, and weighed.

No significant differences were found between the plots A, B and C with respect to root distribution with depth (Kipp, 1981) and the data for these plots have been averaged as shown in Fig. 14a. At D, the pattern was significantly different from those at A, B and C (Fig. 14b).

In all cases, fine root (< 1 mm) concentrations are highest near the soil surface (in the forest floor at plots A, B and C, and in the 0-5 cm mineral soil at plot D), and more or less regularly decrease with depth. At D practically all fine roots are concentrated in the top 20 cm. In the other soils most fine roots are found to 55 cm depth, but appreciable amounts are observed to 80 cm depth. Generally speaking, the coarser roots have their maximum at somewhat greater depth, and, particularly in A, B and C, the coarsest roots (diam. > 1 cm) tend to be concentrated in the lower part of the root zone. The coarser the roots, the greater the spatial variability (Fig. 14). Because the sampling pits were generally at least 1 to 2 meter away from trees, coarse roots, and hence an appreciable part of total root mass, have probably been underestimated. Mean root mass (Table 8) is highest at D, and distinctly lower at A and C. The high variability at B, due to a single thick root found in one of the four columns at 50-55 cm depth, precludes comparison with the other plots.

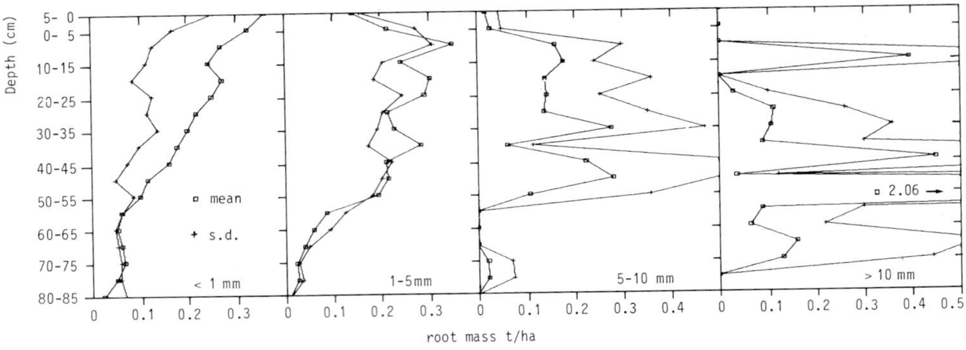

Figure 14a. Profiles of root mass (in t/ha per 5 cm soil layer) of different diameter classes. Means and s.d. for four 25 × 25 cm² soil columns at each of the plots A, B and C (n = 12).

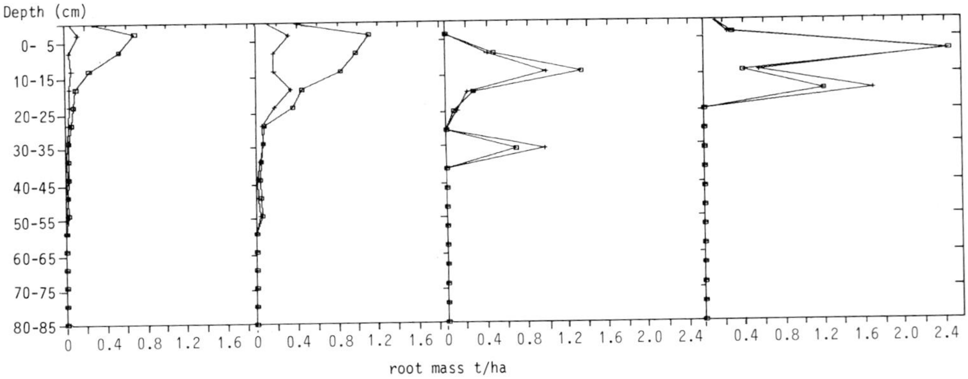

Figure 14b. Profiles of root mass (in t/ha per 5 cm soil layer) of different diameter classes at plot D. Means and s.d. for three 25 × 25 cm² soil columns.

Table 8. Mean root mass (t/ha) down to 85 cm depth in each of the four monitoring plots.

plot	A	B	C	D
mean	7.7	20.0	6.3	26.1
s.d	1.3	16.3	3.5	7.9
n	4	4	4	3

3.2 Changes in vascular plant vegetation over the past decades *(J.P. Kools & A. Ehrenburg)*

3.2.1 Introduction

The changes in environmental conditions in the past decades, including possible strong soil acidification due to atmospheric deposition gave rise to the question: has the vegetation (species composition) changed between 1957/58 (when the first vegetation inventory was made) and 1984? Of the environmental factors that could influence plant growth and species composition we considered light, soil moisture, soil acidity and nutritional state of the soil.

The following hypotheses were formulated:

- Light: Due to tree crown development, we expect a decrease in the quantity of light penetrating to the soil surface, resulting in a decline of plant species typical for sites with high light intensity.
- Moisture state: On account of the regulation of the groundwater regime in the 60's we expect a decline of plant species adapted to wet or moist conditions.
- Acidity: On account of the increasing soil acidification we expect an increase of acid-tolerant species.
- Nutritional state: On account of the large quantities of nitrogen deposited now in the 'Oude Maat' we expect an increase of species of eutrophic stands and a decline of species tolerant of low nitrogen levels.

3.2.2 Methods

Our vegetation survey had to be compared with those by Witkamp (maps of every species recorded for every square 100 m^2, reported by Minderman, 1981) and Bannink & Zonneveld, 1958/59 (typification of the vegetation, some relevés and some single species maps, also reported by Minderman, 1981). Therefore we surveyed every species in every 10×10 m^2 square at the time that the species was most conspicuous (May, June or July 1984), using the scale of Braun-Blanquet after Barkman, Doing and Segal (1964).

3.2.3 Results

For each species we drew a map with the coverage in every square indicated by a symbol (Fig. 15). We compared the data of 1957 and 1984 by comparing the occurrence of species in relation to (1) environmental conditions (2), typification and (3) Ellenberg-indication values.

Occurence of species

We classified every species into one of the following groups: not recorded any-

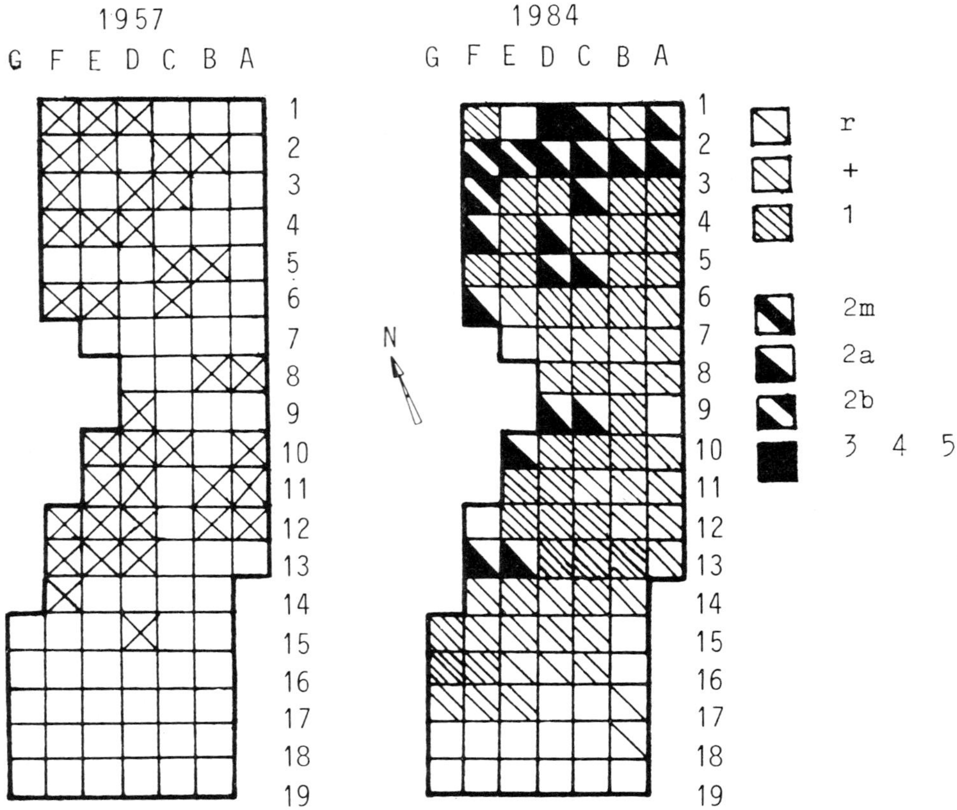

Figure 15. Map showing the presence of *Rubus fruticosus* in 1958 (present in crossed squares) and in 1984 (letter and figure symbols modified after Braun-Blanquet. Increasing shading generally indicates increasing abundance).

more in 1984, decreased, increased, newly recorded in 1984, and more or less the same in both years. We also counted the number of 10 × 10 m² squares in which the species occurred. The result is listed in Table 9. Some species not recorded anymore in 1984 indicate 'rich' conditions (e.g. *Poa pratensis*, *Senecio viscosus*, Agrostis stolonifera), while others indicate 'poor' conditions (e.g. *Festuca ovina*, *Melampyrum pratense*). Some moist-indicating species disappeared (*Ajuga reptans*) whereas others (*Frangula alnus*, *Molinia Caerulea*) increased in abundance. Newly recorded in 1984 are *Prunus serotina* (not indigenous, has increased over the whole country in the last few years), *Galium aparine*, *Poa nemoralis* and *Impatiens noli-tangere*, all of which indicate a high supply of N. The two latter species and *Luzula multiflorum* also indicate more light conditions. Some of the species that have declined indicate poor and acid conditions like *Teucrium scorodonia*, *Holcus mollis*, and *Deschampsia flexuosa*. Because soil acidity certainly has not decreased between 1958 and 1984, the decline of these species is probably a reaction to increased N-supply. However,

Table 9. Occurrence of species in 1957 and 1984 as indicated by the nummer of 10 × 10 m² squares occupied by individuals of the species in question, in 1957, in 1984, and both in 1957 and in 1984.

Not anymore recorded in 1984:	'57	'84	'57+'84
Agrostis stolonifera	93	0	0
Luzula campestris	59	0	0
Antohxanthum odoratum	46	0	0
Hieracium umbellatum	38	0	0
Ajuga reptans	30	0	0
Melampyrum pratense	18	0	0
Cirsium palustre	17	0	0
Hypericum maculatum	17	0	0
Myosotis sylvatica	14	0	0
Poa pratensis	13	0	0
Senecio viscosus	13	0	0
Succisa pratensis	12	0	0
Cardamine pratensis	10	0	0
Poa trivialis	9	0	0
Campanula rotundifolia	8	0	0
Festuca ovina	7	0	0
Equisetum hyemale	5	0	0
Cirsium vulgare	4	0	0
Mycelis muralis	4	0	0
Valeriana officinalis	4	0	0

New in 1984:	'57	'84	'57+'84
Prunus serotina	0	67	0
Agrostis tenuis	0	42	0
Galium Aparine	0	24	0
Circaea lutetiana	0	12	0
Poa nemoralis	0	4	0
Impatiens noli-tangere	0	3	0
Luzula multiflorum	0	3	0

More or less the same:	'57	'84	'57+'84
Lonicera periclymenum	98	98	92
Pteridium aquilinum	91	103	89
Anemone nemorosa	88	79	78
Deschampsia caespitosa	23	25	16
Lysimachia vulgaris	18	19	12
Cornus sanguinea	14	13	11
Urtica dioica	9	10	4
Carex pilulifera	6	4	0
Daphne mezereum	4	3	1

Declined:	'57	'84	'57+'84
Teucrium scorodonia	86	19	18
Viola riviniana	48	7	7
Deschampsia flexuosa	91	59	56
Holos mollis	71	45	40
Galium mollugo	28	2	2
Scropphularia nodosa	28	1	1
Moehringia trinervia	31	16	11
Angelica sylvestris	17	8	6
Veronica chamaedrys	13	2	2
Taraxacum spec.	15	5	3
Listera ovata	9	4	4
Fragaria vesca	7	4	0

Increased:	'57	'84	'57+'84
Sorbus aucuparia	4	99	4
Hedera helix	1	61	1
Rubus idaeus	5	73	5
Frangula alnus	25	80	22
Rubus 'fruticosus'	38	88	37
Galeopsis tetrahit	46	77	30
Polygonatum multiflorum	40	70	39
Molinia caerulea	26	50	24
Maianthemum bifolium	72	92	68
Stellaria holestea	69	86	65
Holcus lanatus	40	53	18
Corylus avellana	6	39	5
Sambucus nigra	1	18	1
Lamium galeobdolon	8	23	8
Crataegus monogyna	9	19	6
Ribus rubrum	3	13	2
Calamagrostis epigejos	1	4	1
Ribes uva-crispa	2	5	0
Viburnum opulus	2	4	2

also species of relatively rich conditions like *Viola riviniana* and *Galium mollugo* have declined.

Species that have increased remarkably are *Sorbus aucuparia*, *Rubus idaeus*, *Rubus fruticosus*, *Frangula alnus* and *Hedera helix*. *Rubus* and *Hedera* indicate a good nutritional status of the soils, just like *Galeopsis tetrahit*, *Polygonatum mutiflorum*, *Corylus avellana* and *Lamium galeobdolon*. However, also species of poor conditions (*Maianthemum bifolium* and *Molinia caerulea*) have increased in abundance. Both light-indicating species (*Corylus avellana, Molinia Caerulea, Gąleopsis tetrahit* and *Stellaria holostea*) and shade-indicating species (*Lamium galeobdolon, Maianthemum bifolium* and *Polygonatum multiflorum*) have become more abundant. *Cornus sanguinea* and *Daphne mezereum*, calcium indicators, and indeed found in the part with calcareous soil, were about equally abundant in both years. Species indicating acid soils as *Lonicera periclymenum*, *Pteridium aquilinum*, *Deschampsia cespitosa* and *L ysimachia vulgaris* have not changed very much either. So, at first sight, we do not observe clear tendencies attributable to the factors light, soil moisture and pH. Vegetation changes clearly indicate an N-enrichment of the soil however.

Comparison between 1958 and 1984

We have made a typification of the vegetation in 1984 and compared it to the one of 1958 made by Bannink and Zonneveld (in: Minderman, 1981). For a brief version of our crosstable see Table 10. The differences between the types 1-4 are rather small, so it is better to speak of local types.

By comparing the different types of 1958 and 1984 it is remarkable that the composition of some types in 1958 has changed so much that they cannot be distinguished anymore in 1984. Examples are the types designated by Bannink and Zonneveld as B1 (many *Teucrium scorodonia*), B3 (with *Populus tremula*) and B4 (with *Corylus avellana*, *Crataegus monogyna, Lamium galeobdolon* and *Cornus sanguinea*). Other types distinguished in 1958 have changed little (in species composition and/or cover) e.g. Cl (1958) and C2 (1984). There are also types in 1984 without a comparable type in 1958; e.g. 4a, b c (with *Corylus avellana* and/or *Lamium galeobdolon*). The changes in typification are not only the result of changes in vegetation (e.g. B1, B4), but also of the fact that every typification is somewhat subjective (e.g. the distinction of 3a, 3b, 3c, 3d). For both years, four main types have been distinguished which can be compared reasonably well (Table 11). These main types have also been used to calculate Ellenberg indication values. The main types indicate an increasing nutrient supply from I to IV. In Table 12 the number of squares per main type in both years is listed.

The spatial distribution of the main types in both years is depicted in Fig. 16. The area of type I has declined in favour of the more eutrophic type II (enrichment of type I mainly with *Rubus idaeus* and *Hedera helix*). One square with type II in 1958 has changed into type III (decline of *Teucrium scorodonia* and increase of *Rubus*

Table 10. Crosstable indicatin species assigned to vegetation types.

species:	vegetation types													
	1a	1b	2	3a	3b	3c	3d	4a	4b	4c	5	6a	6b	7
Pteridium aquilinum	---	---	---	---	---	---	---	---	---	---	---	---	---	---
Sorbus aucuparia	---	---	---	---	---	---	---	---	---	---	---	---	---	---
Lonicera periclymenum	---	---	---	---	---	---	---	---	---	---	---	---	---	---
Stellaria holostea	---	---	---	---	---	---	---	---	---	---	---	---	---	---
Prunus serotina	---	---	---	---	---	---	---	---	---	---	---	---	---	---
Frangula alnus	---	---	---	---	---	---	---	---	---	---	---	---	---	---
Maianthemum bifolium	---	---	-	-	---	-	---	---	-	---	-	---	---	---
Deschampsia flexuosa	---	---	---		---		---	-	-		-	-	-	
Molinia caerulea	---	---	-		---	-	---		-			-	-	
Holcus mollis		---	-	-		---	---		-		-			
Holcus lanatus		---	-	-		---	---		-	---	-			
Anemone nemorosa			---	---	---	---	---	---	---	---	---	---	---	---
Rubus fruticosus			---	---	---	---	---	---	---	---	---	---	---	---
Galeopsis tetrahit				---	---	---	---	---	---	---	---	---	---	---
Rubus idaeus				---	---	---	---	---	---	---	---	---	---	---
Hedera helix				---	---	---	---	---	---	---	---	---	---	---
Polygonatum multiflorum				---	---	---	---	---	---	---	---	---	---	---
Corylus avellana								---	-	---	---	---		
Lamium galeobdolon									---	---	---			
Crataegus monogyna											---	---	---	---
Deschampsia cespitosa											---	---	---	---
Ribes rubrum											---	---	---	---
Galium aparine											---	---	---	---
Populus tremula											---	---	---	---
Circaea lutetiana												---	---	---
Urtica dioica												---	---	---
Viola riviniana												---	---	---
Angelica sylvestris												---	---	---
Daphne mezereum												---	---	---
Galium mollugo												---	---	---

(blank) = occurs rarely
– = occurs regularly
– – – = occurs often

idaeus). In 1984 type IV is only confined to the calcareous area; in 1958 it was also observed in two squares outside the calcareous area. According to the syntaxonomic classification of Westhoff and Den Held (1969), types A1-B4 (1958) can be considered as *Quercion robori-petraeae*, A1 as association *Querco roboris-Betuletum*, and A2-B4 as association *Fago-Quercetum*. Types E1 and E2 can be considered as *Alnopadion*. In 1984 the *Fago-Quercetum* types are still the most important. It can be discussed whether type 3a is still *Querco roboris-Betulum*. In 1984 types 6a, 6b and 7 are *Alnopadion*. From a comparison of main types in 1958 and 1984 we can con-

Table 11. Division of vegetation types in 1958 by Bannink and Zonneveld (in Minderman, 1981) and in this study into main types.

main type	type 1958	type 1984
I	A1, A2	1a, 1b
II	C1, C2, B1	2, 3a t/m 3d
III	B2, B3	4a, 4b, 4c, 5
IV	B4, E1, E2	6a, 6b, 7

Table 12. Number of squares occupied by the main types in 1958 and 1984.

main type	1958	1984
I	41	21
II	19	50
III	34	20
IV	11	14

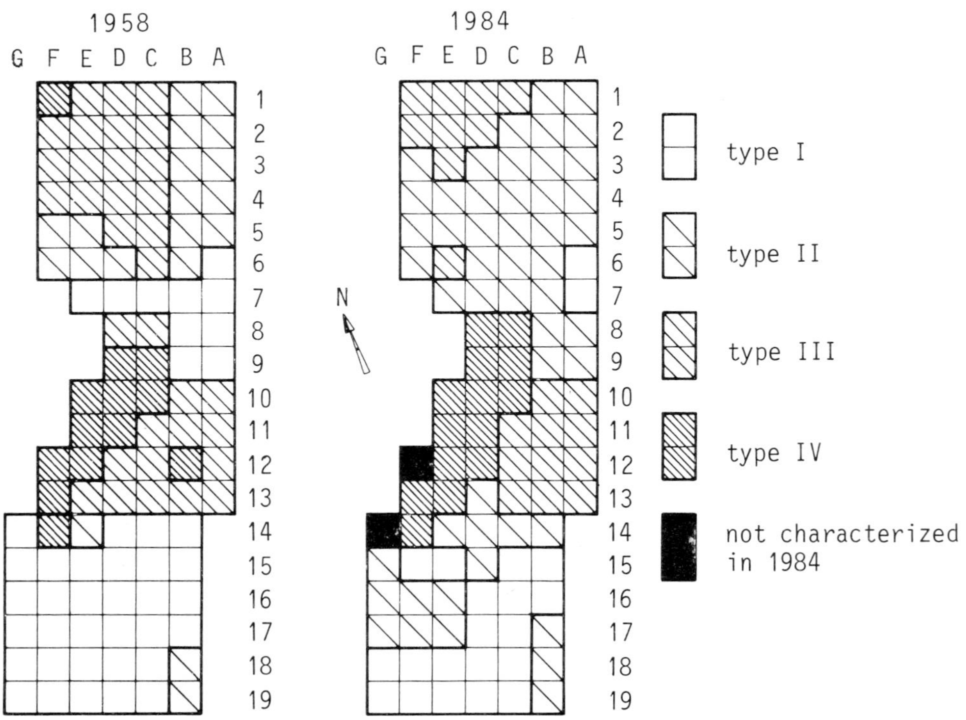

Figure 16. Spatial distribution of the different main vegetation types in 1958 and 1984. Each square is $10 \times 10\ m^2$.

clude only that in general the types have become a little 'richer' and that the types indicating nutrient-poor conditions have declined. In the case of the 'Oude Maat' it is difficult to study changes in vegetation by comparing typifications because the area covered by wood is rather small (3.2 ha) so the types can hardly develop completely. Moreover, the 'normal' development of the ground vegetation under an abandoned coppice is hardly known. Comparing reactions of separate species may therefore be a more fruitful approach.

Both for 1958 and 1984 we have calculated the main indication values (miv's) for light, moisture, pH and nitrogen for every main type, using the indication values Ellenberg (1974) has drawn up for many species. For 1958 we have used frequency (in every main type) as a measure of the importance of early species. For calculating the miv's per factor per main type we used the following formula:

$$\mathrm{miv}_{xI} = \frac{f_{1I}\, e_{1x} + f_{2I}\, e_{2x} + \ldots}{f_{1I} + f_{2I} + \ldots}$$

f_{1I} = frequency of species 1 in main type I (= number of squares (I) with species 1 present/total number of squares (I))

e_{1x} = Ellenberg-indication value of species 1 for factor x

Table 13 shows all calculated miv's. We will briefly discuss the results for the four factors considered:

– Light (L): A slight decline of the miv of this factor in all main types can be seen between 1958 and 1984. This may be the result of development of the tree canopies. In recent years, however, canopy closure may have decreased.

– Soil moisture (F): the miv's for soil moisture are constant or have increased very slightly. So the measures to improve drainage in the 60's have not greatly affected the species composition.

– pH (R): There is no evidence for a change in vegetation due to decreased soil pH between 1958 and 1984. The expected differences among the four main types, I (most acidic) to IV (calcareous) were confirmed, however.

– Nitrogen (N): Both for 1958 and 1984 the miv's for nitrogen increase when going from main type I via II and III to IV as expected with increasing soil nutrient levels. The expected tendency of the miv for nitrogen of every type to increase between 1958 and 1984 was also confirmed, suggesting that atmospheric deposition of N has led to vegetational changes.

3.2.4 Conclusions

Of the expected trends in species composition due to (1) decreased light intensity, (2) decreased soil wetness, (3) increased soil acidity and (4) increased N-availability only (1) and (4) could be confirmed. Decreased soil wetness is probably mainly reflected by shorter periods of very high groundwater levels in winter and early spring,

Table 13. Mean Ellenberg-indication values for light (L), moisture (F), acidity (R) and nitrogen (N) per main type in 1958 and 1984, based on species frequency.

main type	factor	1958	1984
I	L	6.0	5.7
	F	5.1	5.5
	R	3.3	3.2
	N	3.4	3.6
II	L	6.0	5.4
	F	5.1	5.3
	R	3.6	3.8
	N	3.8	4.4
III	L	5.9	5.5
	F	5.2	5.3
	R	4.0	4.2
	N	4.3	4.6
IV	L	6.1	5.7
	F	5.4	5.4
	R	4.8	5.3
	N	4.9	5.5

while particularly between May and December soil moisture conditions have probably changed little.

In view of the strong acidification of the surface soil due to the very high $(NH_4)_2SO_4$ inputs (which must be a recent phenomenon) the pH of the mineral surface soil has almost surely decreased over the past decades. That this has not clearly influenced the ground vegetation may be due to the fact that many species root mainly in the litter layer.

It should be emphasized here that this discussion deals with changes in vegetation and correlated changes in environmental factors, but that causal relations between vegetation and environment cannot be revealed in this study.

3.3 Changes in fungus flora between 1957-58 and 1982-84 *(A.E. Jansen & B.W.L. de Vries)*

3.3.1 Introduction

The research on soil biology in 1957 and 1958 included a study on the fungal species and the growth of mycelium in litter and humus in three plots of different humus type: calcareous mull, acid mull, and mor (Witkamp, 1960; Minderman, 1981).

In 1982, 1983 and 1984 these 3 plots in Hackfort were investigated again for fungal fruitbodies, to see whether the fungus flora has changed since 1957-58. This study is part of the ecological-mycological research program of the Biological Station at Wijster. An important part of this research consists of monitoring of permanent plots, and comparing recent and old data. This research has indicated a considerable change in mycoflora, in particular a strong decline in mycorrhizal fungi, in the last 25-50 years throughout the Netherlands (Arnolds, 1985).

3.3.2 Fungi in 1982-1984

Methods

Three plots, situated where Minderman did his research, were investigated (Fig. 17).

Plot 1 contains the whole area that we could clearly classify as calcareous mull, and comprises squares D10, plus parts of C10, E10, C11, D11, E11, D12 and E12. The surface area is about 400 m^2, considerably less than the 800 m^2 calcareous mull mentioned by Witkamp. The green vegetation belongs to the *Alno-Padion.*

Plot 2 comprises squares D2, D3, E2, E3, F2 and F3, an area of 600 m^2, adjacent to monitoring plot A. Humus is of the mor type, but was an acid mull type 25 years

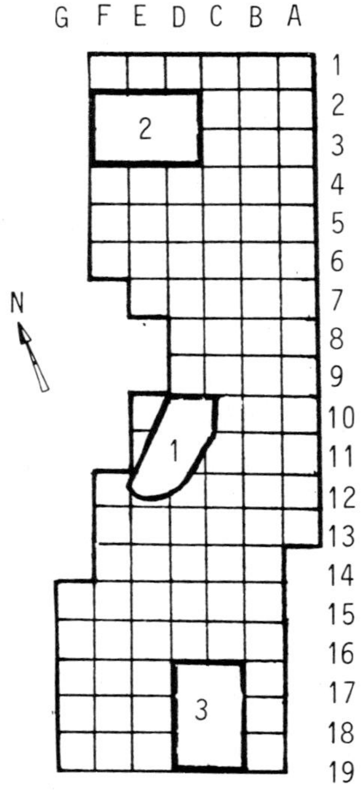

Figure 17. Location of the plots used in studying the fungus flora.

ago. The green vegetation is intermediate between *Querco-Carpinetum* and *Violo-Quercetum* (= *Fago-Quercetum).*

Plot 3 comprises the squares C17, C18, C19, D17, D18 and D19 with a surface area of 600 m^2, adjacent to monitoring plot B. Humus is of the mor type; the green vegetation belongs to the *Querco-Betuletum.*

The plots were visited in 1982 on 14 October (relevés 1-3) and 10 November (relevés 4-6), in 1983 on 13 October (relevés 7-9) and 1 November (relevés 10-12), in 1984 on 24 October (relevés 13-15). Relevés are given per plot, species are alphabetical (Appendix 4). The fruitbodies of macrofungi were counted. Because of their small size, fruitbodies of microfungi were counted, but their abundance was estimated or their presence was indicated. Specimens are kept at the herbarium of the Biological Station at Wijster (see Appendix 4D).

The similarity or affinity of the plots in terms of fungus flora of the plots was calculated according to Barkman (see e.g. Jansen, 1981):

$$S = c + 1 / [(a+b)^{1/2} + 1] \qquad (1)$$

(S = similarity, c = number of species in common, a = number of species occurring in one plot only, b = number of species occurring in the other plot only).

Results

In total 156 species have been found (Appendix 4). They belong to the taxonomical groups of the *Agaricales* (95 species), *Gasteromycetes* (3 species), *Aphyllophorales* (41 species), *Ascomycetes* (16 species) and *Myxomycetes* (1 species). All Agaricales and Gasteromycetes and the species with large or distinct fruitbodies of the other taxonomical groups are called 'macro-fungi' (114 species). Species with small fruitbodies of the Aphyllophorales, Ascomycetes and Myxomycetes are called 'micro-fungi' (42 species).

Plot 1 appeared to be much richer in species than the other two plots. (Table 14), viz. 103 species, compared to about 70 for the other two plots. The species number in plot 1 is very high, especially so because the area of this plot is smaller than of plots 2 and 3. Plots 2 and 3 have about the same number of species as 1000 m^2 plots in oakwoods of the same vegetation type in Drenthe (Jansen, 1981).

Table 14. Number of species per plot in 1982-1984, in parentheses the number of species of macrofungi.

plot	1	2	3
total	103(67)	74(59)	69(46)
observed in all 3 plots	28(24)	28(24)	28(24)
observed in 2 plots	8(6)	8(6)	–
	–	19(13)	19(13)
	7(3)	–	7(3)
observed in 1 plot only	60(43)	19(15)	15(6)

The similarity of the 3 plots in terms of the fungus flora, calculated according to (1), is as follows (parentheses refer to macrofungi only):
plot 1 – 2: S = 0,7 (0,8)
plot 1 – 3: S = 0,7 (0,9)
plot 2 – 3: S = 1,9 (2,6)

So the fungus flora (especially macrofungi) in plots 2 and 3 are closely related to each other, and much less to that in plot 1.

3.3.3 Fungi in 1957-1958

Methods

Data on the fungus flora in 1957 and 1958 were collected by M. Witkamp (Witkamp, 1960). In both years he visited the plots 13 times. The areas he considered were 800 m^2 for calcareous mull, 6600 m^2 for acid mull and 1600 m^2 for mor humus. He did not count the number of fruitbodies; the numbers in the table are the numbers of visits that a species was present. Minderman (1981) published again Witkamps observations from 1957. I compiled both lists (Appendix 4B, species names according to Arnolds, 1984).

Results and discussion

Mindermans list contains 25 species (14 terrestrial, 11 growing on wood) not mentioned by Witkamp. Including those 25 species, 75 species were reported for 1957-58 (Table 15).

Plot 3 (mor humus) is most rich in species (50) followed by plot 2 (acid mull) with 46 species; plot 1 (calcareous mull) is poorest with 41 species. The similarity of fungus flora calculated according to (1) is as follows:
plot 1 – 2: S = 2.0
plot 1 – 3: S = 1.6
plot 2 – 3: S = 1.0

Table 15. Number of species per plot in 1957-1958.

plot	1 calcareous mull	2 acid mull	3 mor
total	41	46	50
observed in all 3 plots	22	22	22
observed in 2 plots only	7	7	–
	–	10	10
	2	–	2
observed in 1 plot only	10	7	16

So, the fungus flora of plot nrs. 2 and 3 are closer related to that in plot nr. 1 than to each other.

Even though in 1957-58 the plots were studied more frequently than in 1982-84, the number of species found in 1957-58 is considerably smaller than in 1982-84 even if we compare only the numbers of macrofungi. However, the species lists of 1957-58 are incomplete, because 1) only 2 autumns were studies and most fungal species do not occur each year (e.g. Jansen, 1981); 2) Minderman studied only the terrestrial species, and 3) all the small Agaricales were probably overlooked (no small Agaricales were reported but it is very unlikely that they did not occur). Identification of about 25 species is questionable, because of their quite different ecological preference or because of recent taxonomical views (see Appendix 4B). The identification could not be checked, as no specimens were conserved.

3.3.4 *The fungus flora in 1957-1958 and 1982-1984 compared*

The species of each plot were divided into 3 groups: the species 'reported again', 'not reported again' and 'newly found' (see Appendix 4C). The numbers in these groups are as follows (newly found species split in macrofungi and microfungi):

plot	1	2	3	aggregate
reported again	14	15	13	26
not reported again	27	31	37	48
new: macrofungi	62	43	32	88
microfungi	26	16	22	42

Although the recent plots are much smaller than the old ones, the total number of macrofungi observed in 1982-84, 114, is 1.54 times the old number. For the plots separately, these factors are 1.85 in plot 1, 1.26 in plot 2, and 1.10 in plot 3. The total number of species in 1982-84 (156) is twice as much as in 1957-58 (74). About the same number of species were reported again for each of the plots. But these numbers are small in comparison with the numbers of species not found again. In plot 1 66% of the species present in 1957-58 were not observed in 1982-84, in plot 2 67%, and in plot 3 even 74% (3 plots aggregated 65%). So the proportion of species reported again is smallest in plot 3 and largest in plot 1. 130 species are newly reported. Microfungi are important (32%) among the newly found species. As microfungi were not included in the 1957-58 research, we also excluded them in the comparison below.

Three ecological groups of macrofungi can be distinguished:

1. mycorrhizal fungi,
2. litter and humus saprophytes and
3. wood saprophytes and parasites.

Table 16 shows that the increase in number of fungal species in plot 1 and 2 is caused only by an increase in saprophytes. The aggregated number of saprophytes

Table 16. The number of species of macrofungi 'new', 'reported again' and 'not reported again' in the ecological groups of mycorrhizal, litter and humus, and wood fungi.

	not reported again	reported again	new
mycorrhizal fungi			
plot 1	12	3	9
2	21	2	8
3	23	3	6
aggregate	30	7	12
litter and humus fungi			
plot 1	14	6	42
2	5	8	27
3	10	6	18
aggregate	15	10	62
wood fungi			
plot 1	1	5	11
2	5	5	8
3	4	4	8
aggregate	3	9	14

(groups 2 and 3) for 1982-84 is a factor 2.57 larger than for 1957-58. For the plots separately, these factors are 2.45 in plot 1, 2.09 in plot 2 and 1.50 in plot 3. The mycorrhizal fungi, on the contrary, decreased strongly: the aggregated number for 1982-84 is only 51% of the number for 1957-58 (plot 1: 80%, plot 2: 43%, plot 3: 35%). In plot 3 mycorrhizal fungi decreased most strongly and saprophytes increased least. Because the decrease in mycorrhizal species is smaller than the increase in saprophytes, the recent number of species in plot 3 is smaller than the old number.

The similarities of the 1982-84 fungus flora with those of 1957-58, calculated according to (1), gives the following results:
plot 1: S = 0.36
plot 2: S = 0.43
plot 3: S = 0.39
The similarities are very small, indicating that the fungus flora have changed considerably; changes are biggest in plot 1, slightly smaller in plot 3 and smallest in plot 2.

Plot 3, on mor humus, an acid and relatively nutrient poor soil, shows a loss of species. The two other plots, which are less nutrient poor, do not show a loss in number of species. A loss of species is also known from nutrient poor vegetations like the *Dicrano-Quercetum* stands and coniferous forests in Drenthe (Arnolds, 1985).

Formerly fungus flora of the 3 areas studied, had little affinity. Fungus flora of all plots changed, those of plots 2 and 3 became more related to each other and less to that of plot 1. As plot 1 covers only a very small part of and plots 2 and 3 are representatives of a much larger area, diversity of the fungus flora in 'Hackfort' decreased in the last 25 years.

3.3.5 Possible causes of changes in fungus flora

Changes in fungi may be caused by changes in a. vegetation, b. litter layer and humus type, c. in soil drainage, and d. atmospheric deposition.

a. Changes in vegetation

A comparison of vegetation relevés of 1958 with data from 1984 shows that vegetation has changed as a result of eutrophication and that there is a tendency to a more uniform vegetation (Kools & Ehrenburg, chapter 3.2). They attributed changes in herbaceous vegetation mainly to increased N-availability and decreased light intensity. More stems and thick branches rotting away may result in a habitat for several fungal species, e.g. *Piptoporus betulinus*, *Ganoderma applanatum, Gymnopilus penetrans* (all 3 newly recorded). Canopy closure may have resulted in a more humid microclimate, which is a condition for fructification of thin-fleshed fungi, e.g. *Marasmius epiphylloides*, *Mycena stylobates* and *Pleurotellus herbarum* (all 3 newly recorded).

b. Changes in litter layer and humus type

A closed canopy will also result in an increased litterfall, which may stimulate many litter dwelling species from genera as *Mycena*, *Clitocybe*, Psathyrella. Of these genera indeed many species are newly found. A thick litter and humus layer is apparently unfavourable for a number of mycorrhizal fungi, e.g. *Amanita fulva* (= *A. vaginata* in Appendix 3B), *A. muscaria*, *Cortinarius hemitrichus*, *C. paleaceus*, *Lactarius chrysorrheus*, *Russula cyanoxantha*, *R. fragilis*, *R. laurocerasi*, *R. nigricans* and *Tricholoma columbetta.* These are species mainly found on nutrient- and humus-poor soils, e.g. on fixed inland sand dunes (Jansen, 1981). Increased litterfall from *Populus* trees planted recently west of the wood favoured *Hymenoscyphus immutabilis* (a saprophyte on populus leaves) and possibly too the three *Typhyla* species (saprophytes on populus leaves or on nitrogen-rich humus). The change in the fungusflora of plot 2 can be attributed to the change in humus type, from acid mull to mor. This change may explain why the plot 2 fungus flora became more related to that of plot 3 (similarity increased from 1.0 to 2.6).

c. Changes in soil drainage

Groundwater is regulated and since the late fifties the land is not inundated any more in winter. Although in summer drought is probably not more intensive than in the fifties, increased drought in spring is a possible stress here, especially in plots 2 and 3. Mycorrhizal fungi are known to be more sensitive to drought stresses than other fungal species (Mexal and Reid, 1973).

d. Changes in atmospheric deposition

Increased ammonium input may hamper mycorrhiza formation (North American Conference on Mycorrhiza, 1981). Of the species disappeared (Appendix 4C) 62% is mycorrhizal, of the newly found species only 13% is mycorrhizal. Increased N-availability might be the cause of the (new) appearance of species of *Entoloma* and *Naucoria* and of the disappearance of *Marasmius androsaceus*. Some species that disappeared from this wood became much rarer in the whole of the Netherlands, e.g. *Amanita citrina*, *A. phalloides*, *Cortinarius alboviolaceus, C. elatior*, Dermocybe cinnabarina, *Inocybe geophylla*, *Lactarius blennius*, *L. chrysorrheus*, *L. vellereus* , *L. vietus*, *Russula cyanoxantha*, *R. fragilis*, *Tricholoma sulphureum*, *Xerocomus subtomentosus*. Changes in soil pH and nutritional supply due to atmospheric deposition are seen as the main cause of the decrease of these fungal species (Arnolds, 1985). Decreased availability of Ca^{2+}, as a result of increased soil acidity might have contributed to the disappearance of *Calocybe ionides*, a calciphilous species, of *Amanita phalloides*, also calciphilous, and to the disappearance of *Dermocybe cinnabarina*, and some *Lactarius* and *Russula* species.

The reaction of a fungal species to atmospheric deposition, to decreased inundation, to changes in humus and vegetation type depends on its ecological preference and its ecological niche. Knowledge on these ecological niches is still limited, so it is difficult to give the exact cause of changes for each fungal species separately.

3.3.6 Conclusions

In 1982, 1983 and 1984 156 species of fungi belonging to the *Agaricales*, *Gasteromycetes*, *Aphyllophorales*, *Ascomycetes* and *Myxomycetes* were observed in 3 plots: 103 species in plot 1 (calcareous mull), 74 species in plot 2 (mor humus formerly acid mull) and 69 species in plot 3 (mor humus). Data from 1957 and 1958 refer only to the larger 'humus inhabiting' fungi and those on growing on wood were excluded. Moreover, identification of some species is doubtful. Only 26 species were found back in 1982-84; 48 species were not found again, and 130 species were newly reported in 1982-84. Mycorrhizal species decreased. In plot 1 and 2 the increase of saprophytical fungi is larger than the decrease of mycorrhizal fungi. In plot 3, however, saprophytes increased less than mycorrhizal fungi decreased, so the total number of macrofungi has decreased.

These changes in fungus flora are probably caused by changes in vegetation and humus type, by better soil drainage and by increased atmospheric deposition. The cause of disappearance or appearance can be different for each fungal species. The decrease of the mycorrhizal species observed here is a general phenomenon in the country, and is generally atributed to increased atmospheric deposition.

Part 2

Monitoring programme

4 Field and laboratory methods

Figure 18 gives an impression about the installations used. The detailed maps shown in Figure 19 give the exact locations of the various equipment used in the monitoring work. The equipment and methods will be described in detail below.

4.1 Above-ground hydrology *(N. van Breemen & Th. Pape)*

Precipitation was monitored continuously with a Hellman-type raingauge (collection surface 200 cm^2 at 1.2 m above the soil surface), situated about 400 m west of the woodland area. The one-week recorder strip of 80 mm width corresponds to 10 mm of precipitation (manufacturer Lambrecht, Göttingen). The temperature in the recorderhouse was kept above 15°C. No separate snow collectors were used. Daily precipitation values read from the recorder strips are shown in the data files (micro fiche).

Precipitation was also measured fortnightly in the sample collectors for chemical rainwater analysis, placed in pasture land, resp. about 100 m West and 100 m East of the woodland area. The collectors consisted of 400-cm^2 black high-pressure polythene funnels fitted directly on top of 5-liter opaque polythene bottles. To prevent birds from perching on funnel edges, the funnels were supplied with pointed crowns cut out of 2 mm thick polythene sheet. Initially the collectors were placed at about 120 cm above the land surface. To minimize the risk of contamination of the collector as a result of agricultural activities (grazing, sprinkler irrigation, fertilizing and manuring) they were placed at 400 cm above the land surface after June 1981.

Throughfall was collected in 400-cm^2 black high pressure polythene funnels (with pointed crowns against birds), fitted on 5-liter opaque polyethene bottles. Per plot seven collectors were installed, roughly in a hexagonal arrangement (see Fig. 19) with the hexagon placed at random in the plot. The bottles were slid into vertically placed PVC pipes of 20 cm diameter, with the rim of the funnel at about 120 cm above the land surface. From mid 1983 onwards a cone-shaped nylon screen (mesh size about 0.5 mm) was placed at the bottom of the funnel to prevent coarse vegetation particles and insects from dropping into the bottles.

From April to December 1983, throughfall of the herbaceous vegetation was collected in rectangular open-top PVC boxes (6 cm high, 5×20 cm^2 surface area) with a half-circular 5×20 cm^2 trough, perforated with five 0.5 cm ϕ diameter holes in the bottom, sunk partly into the box. Six collectors were used in each plot.

Throughfall quantity was read in the field with a measuring cylinder. If through-

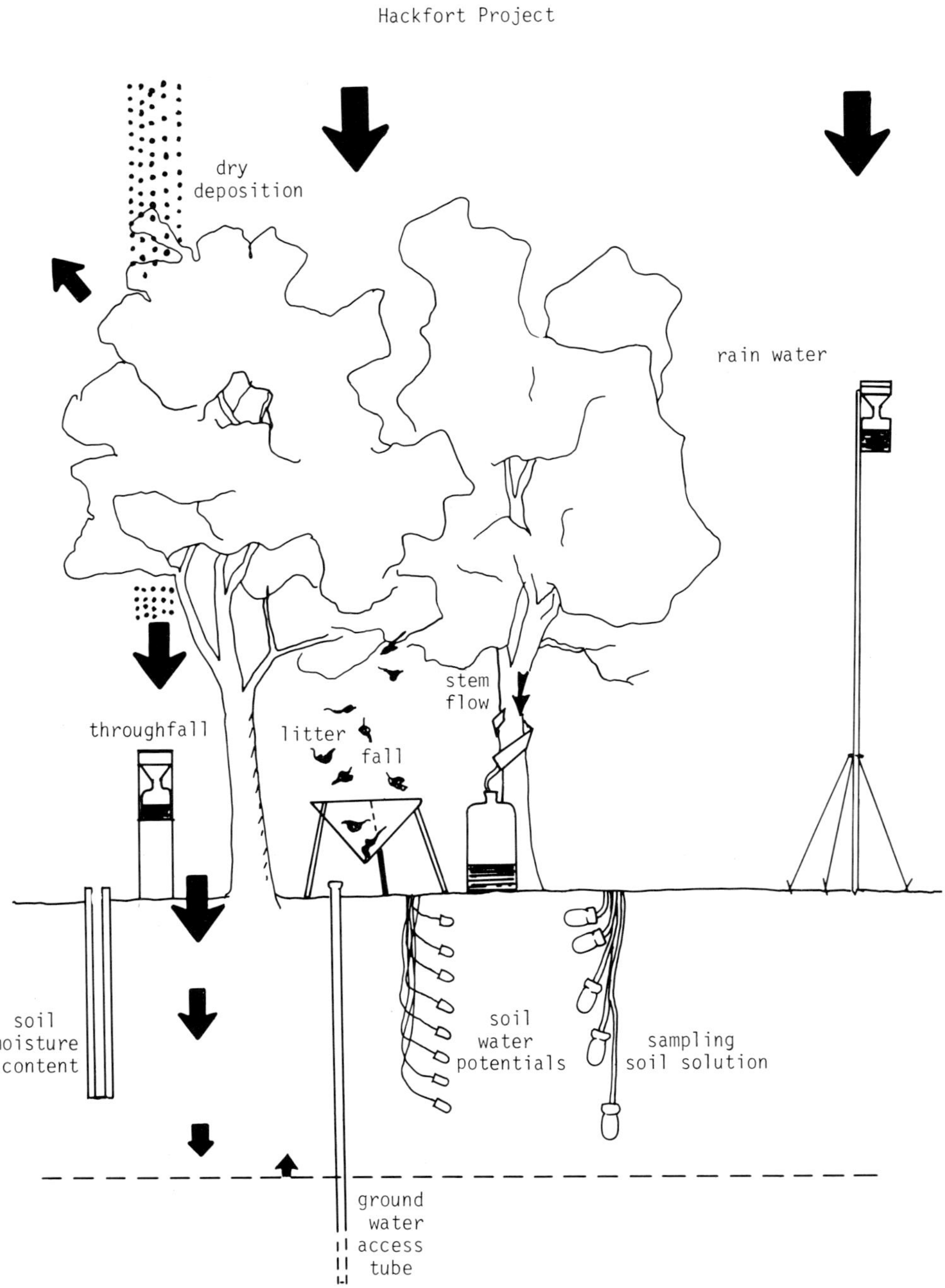

Figure 18. Schematic vertical cross section of the woodland ecosystem and neighbouring open area, with the equipment used in the monitoring programme. Black arrows indicate the flow of water through the atmosphere and through the unsaturated soil.

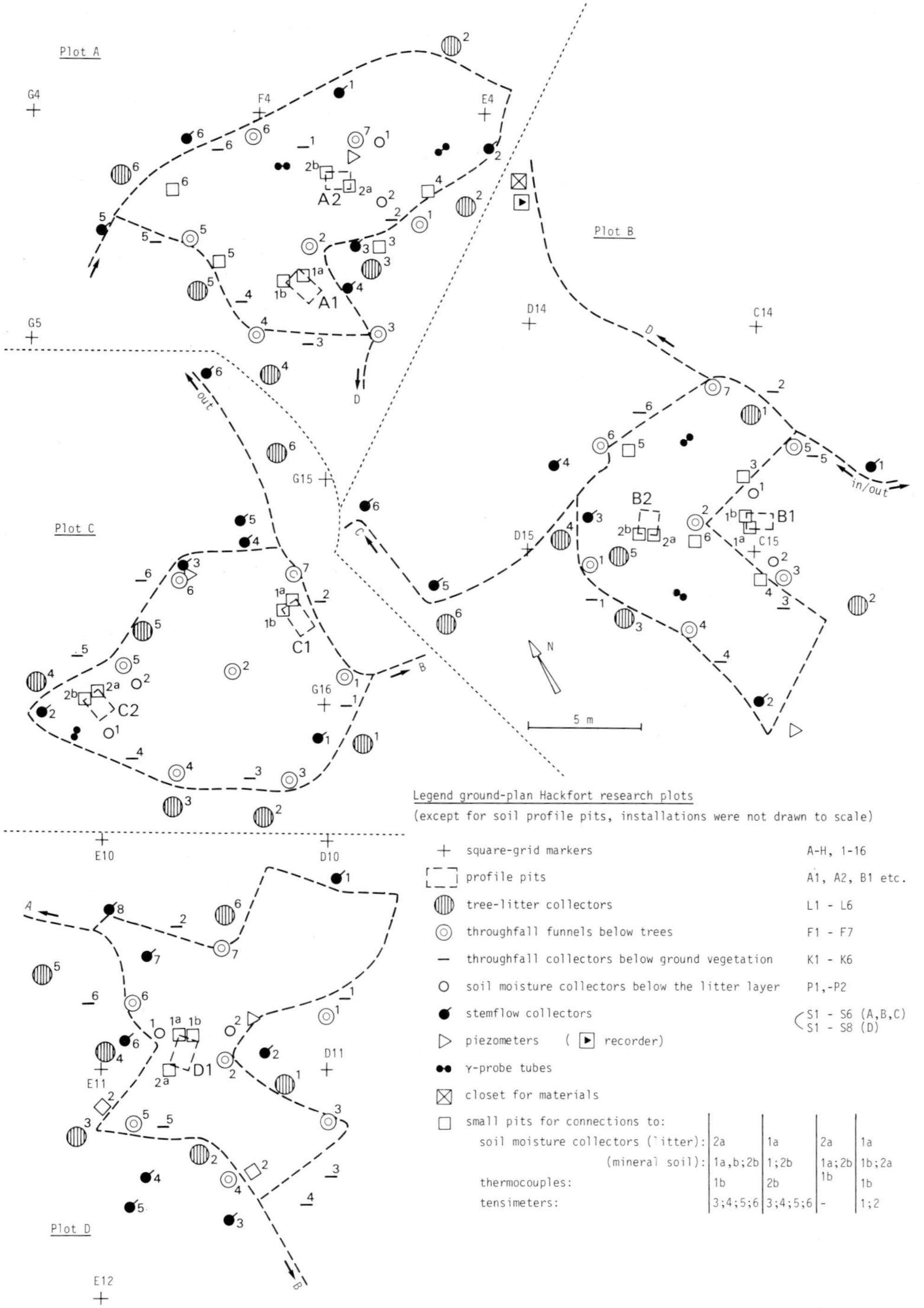

Figure 19. Detailed map showing the locations of the various installations used in the monitoring programme. The broken line indicates the foot path used during the monitoring work.

fall water was frozen, the bottles were replaced by empty ones, and brought to the lab.

Stemflow was collected by spiral collars, encircling trees for about 400°, made of about 10 cm wide, 1 mm thick polythene strips fitted to the trees by screws and silicone kit. Water drained through a nearly vertical hole at the bottom end of the collar through a 6-mm polythene tubing into a 62-liter polythene bottle. The white polythene bottle was painted dark green to minimize algal growth. The quantity of the stemflow was measured fortnightly by means of a calibrated measuring 'blow' tube (stainless steel) that was lowered into the bottle from the top. Six to eight trees at each plot were chosen at random for stemflow collection. Species names and tree diameters at the height of the collars are given in Table 17.

4.2 Soil physical monitoring *(J.J.M. van Grinsven)*

Soil temperature was measured weekly (from August 1980 to January 1981), fortnightly (from January 1981 to January 1982) or monthly (after January 1982, simultaneously with collection of soil solutions) by means of copper-constantan thermocouples, permanently installed at 5, 10, 20, 50 and 100 cm depth, and a Comark type 1624 Electric Thermometer, using an ice-water mixture as reference. One set of five thermocouples was placed in each plot.

Groundwater depth was measured at 6 sites by means of piezometers with a tube diameter of 4 cm and a perforated part of 25 cm length at depths 180 to 220 cm below the soil surface, i.e. well below the lowest groundwater level. During the instal-

Table 17. Diameter of trees (cm) used for collection of stemflow (diameters measured in November 1984, at 150 cm above the land surface). Trees are Quercus robur except if noted otherwise.

stemflow collector nr.	plot			
	A	B	C	D
1	15.5	20	24.5	13.5; 24; 16[a]
2	14.5	18.5	12	13.5
3	17.5	29.5	13	24
4	21.5	33	19.5	30.5[b]
5	22	15	29	10
6	15.5	21.5	13.5	15.5; 13.5; 9.5[c]
7	–	–	–	9[d]
8	–	–	–	11

a Betula (3 stems from one stool)
b Populus
c Alnus (3 stems from one stool)
d Alnus

lation of the piezometers the filters (~ 1 mm wide slits, perforated part surrounded by nylon screen) were embedded in fine gravel in order to increase permeability. One piezometer was installed at each plot. Readings were taken weekly (before spring 1984) or fortnightly (since spring 1984) by means of a plummet with a light signal. One piezometer just north of plot B (see Fig. 19) was monitored continuously by a Stevens model 7003 groundwater recorder. The recorder was calibrated by comparison with weekly readings from a reference piezometer at 1.5 m distance from the recorder. From October 1980 until August 1981 readings were taken hourly. The recording was interrupted several times due to instrument failures. Response time of the piezometers was checked by measuring the time needed to restore the original groundwater level after raising the water level in the piezometer by 10 cm. At plots A, B, and C, the original levels were restored within 5 minutes. After one hour the piezometer level was still 2 cm above the original level in D and 7 cm above the original level in the reference piezometer.

Soil mosture potentials (hydraulic heads) were measured by means of ceramic tensiometers, using a pressure transducer (NS LX 1604 GB). The home-made tensiometer cups were 6 cm long and 1 cm wide, and equipped with double access copper tubing to allow air removal during flushing.

Tensiometer sets at plot A and B were installed from a single ~ 1 m wide trench, through augerholes, horizontally (for tensiometers at 10 and 20 cm depth), or under an angle of about 45° (for deeper tensiometers). Very fine sand (main fraction between 35 and 75 μm) was flushed to the bottom of the hole and around the cups to improve hydraulic contact. Installation under an angle was chosen to prevent disturbance of soil above the tensiometers and to minimize preferential vertical infiltration down the former augerholes. Yet, sometimes positive pressure heads were observed after periods of very high precipitation, indicating that the arrangement chosen might in fact have enhanced preferential infiltration to the cups.

At both plot A and B, sets of tensiometers were installed; at 2 sites from 10 to 90 cm depth and at 2 sites from 60 to 90 cm depth, all at 10 cm depth intervals. As reference point for the hydraulic head measurements, zero hydraulic head at the soil surface was taken. Hydraulic heads were read from April until November with different intervals in different years; fortnightly in 1981, weekly in 1982, and monthly in 1983. No readings were taken in 1984.

At plot D tensiometers were installed (at 2 sites) in spring 1983, at 10, 30 and 50 cm depth. These tensiometers were installed horizontally from a small access pit. They were read monthly in 1983.

During dry periods, tensiometers were flushed after each measurement. In wet periods, they were only flushed when there was evidence for large air pockets inside the tubing or cup. When connecting the pressure transducer to the tensiometer outlets, air inclusion was minimized by preflushing the tensiometer outlets and by maintaining a water flow from the outlet of the pressure transducer during connection. Multivalves were used to connect several tensiometers simultaneously. Readings were generally stable within a few minutes. Fast stabilisation of the reading was also favoured by a relatively small diurnal temperature variability due to shade from trees. Measured hydraulic heads never exceeded 800 cm, which is the air entry value for the cups. We consider the readings to be accurate within a few cm near saturation, and within 10 to 20 cm near the air entry value.

From June 1983 to November 1983, hydraulic heads were monitored as a function of distance to a tree using five transects of three tensiometers from one central tree to five neighbouring trees. The three tensiometers were installed at 30 cm depth, at one half, one fourth and one eighth of the distance between the central tree and the neighbouring tree, starting from the central tree. Pressure heads were read fortnightly by means of manometers (Jet-fill tensiometers; Soil Moisture Equipment Corp.).

Soil water content was measured by the gamma attenuation method using a TROXLER, model 2600 double access tube gamma scintillation probe. At plot A and B tubes were installed at 2 sites, at plot C at 1 site. During installation guide rings were used to assure tube alignment. Water contents were read monthly (with variable winterstops) between 1980 and 1984 at 2 inch depth-intervals (0.051 m) down to a depth of 36 inches (0.914 m). The gamma probe was calibrated twice. A first preliminary calibration consisted of gravimetric sampling of soil from the access holes for the tubes, followed by a gamma scan, immediately after installation of the tubes. This calibration was rather poor for the topsoil: the very low bulk density of the soil made it almost impossible to sample it without changing soil bulk density. The calibration curves were calculated by smoothing the gravimetrically determined moisture content profiles. A second calibration was carried out in 1985. After a gamma scan two undisturbed 100 cm^3 soil cores were sampled at 10 cm depth intervals at 20 cm from the access tubes. Both calibrations gave comparable results but in the second calibration replicate moisture contents agreed far better.

4.3 Water sampling and analysis *(N. van Breemen & E.J. Velthorst)*

All water samples were collected and stored in polythene containers. Containers were cleaned only by rinsing with distilled water; the same container was always used for a sample from a specific collector.

Soil solution in the mineral soil was collected by means of type 1910 high flow porous ceramic cups (Soil Moisture Corp., Santa Barbara, California) at 10, 20, 40, 60 and 90 cm depth. Duplicate cups were placed at 10, 40 and 90 cm, and after September 1983 four cups were used at 90 cm depth. Before installation, the cups were percolated with 0.001 M HNO_3 followed by prolonged leaching with demineralized water. Each cup (a, see Fig. 20) was fitted with a perforated silicone rubber bung (b), which in turn was connected to a nylon-reinforced polythene tube (c, 8 mm internal ϕ), closed by a small rubber stopper (d) with two access tubes of tygon. One of the access tubes (f) could be used to evacuate the porous cup, the other (e) to sample the water collected inside the cup. The cups were installed through a vertical augerhole to the desired depth. Soil from the augerhole was placed around and abo-

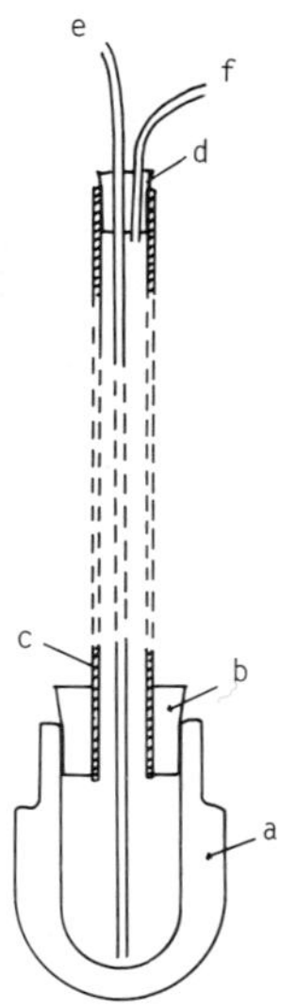

Figure 20. Cross section diagram of porous ceramic cups used to collect soil solutions in the mineral soil (not to scale). For explanation see text.

ve the cup restoring the original soil profile (as much as possible) and compacted firmly by a wooden stick. Suction was applied either by evacuating the cups with a hand-operated vacuum pump (when the soil was very wet) or by connecting the cups to a ±4 liter PVC tank previously evacuated in the laboratory (when the soil was moist to dry).

Within 24 hours after evacuating the cups, the water collected in the cups (usually 10 ato 100 cm^3) was sampled through tube *e,* by suction into a 100-cm^3 polyethylene bottle and brought to the laboratory. Soil solutions were sampled monthly except in frost periods. After prolonged rainless periods in summer, the surface (0-60 cm) soil was often too dry to collect water.

The soil solution at 0 cm depth was sampled by means of a filter plate (two per plot) placed immediately below the organic surface layer (Fig. 21). The filter plate consists of Acropor filter (0.2μm pore diatemeter, Gellmann Inc) (a), over a circular 19 mm thick body of porous polythene (Supralen RCH 1000, Mannesmann GmbH) (b), into a circular PVC ring (c) (internal ϕ 10 cm) with a PVC bottom plate (d). Suction was supplied continuously by means of a siphon from a water-filled bottle (f) to the groundwater in a piezometer tube (g), and sample was collected in polythene bottle (e). The sample bottle was placed under the soil surface in the dark. The soil solution at 0 cm depth was collected fortnightly and, after measuring the quantity (in mm) and storage at 4° C, pooled to monthly samples for chemical analysis. The organic surface layer inside the PVC ring contained no living roots, so that the percolate collected is influenced by mineralization of above-ground tree litter, without the effect of assimilation by plants.

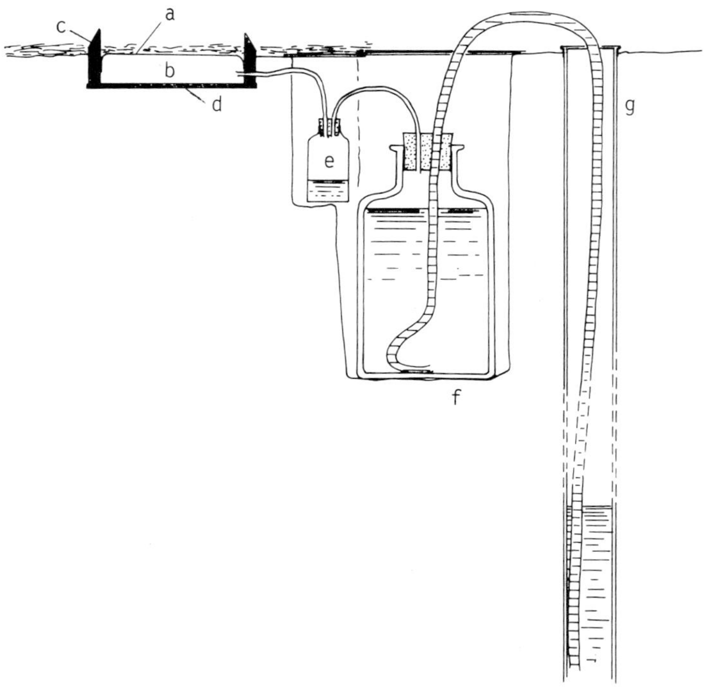

Figure 21. Cross section diagram of the porous plate and the hanging drop arrangement to collect soil solutions at the bottem of the forest floor. For explanation see text.

Ground water was collected monthly from the piezometers by means of a PVC pipe with a stainless steel valve at the bottom.

Measurements of the quantities of rain, throughfall and stemflow water started in 1980, routine chemical analysis started only in 1981. Bottles for collection of rain water were replaced fortnightly and brought to the laboratory for measuring the quantity and filtering before analysis. In 1981 fortnightly samples were analyzed. Later mid-monthly samples were stored at 4°C before pooling them to monthly samples for analysis. Sample volumes of throughfall in individual collectors were measured fortnightly in the field. The throughfall bottles were rinsed with demineralized water and were replaced only if the samples were frozen.

From January 1981 up to January 1982 (for plot A, B and C) or December 1982 (for D), samples were pooled in pairs (from collectors 1 & 2, 3 & 4 and 5 & 6; samples from collector 7 were discarded). Later, monthly samples from all seven collectors per plot were combined to one volume-weighted sample.

Stemflow samples from individual collectors at each plot were combined to volume-weighted pooled samples; fortnightly and in pairs (1 & 2, 3 & 4, 5 & 6) in 1981, fortnightly in threes (1 & 2 & 3; 4 & 5 & 6) in 1982. From 1983 onwards, samples from all collectors per plot were pooled to one sample. Measuring and analysis of stemflow was stopped in January 1984.

Throughfall of the herbaceous vegetation was collected fortnightly and pooled to one volume-weighted monthly sample per plot.

Rainwater, throughfall, stemflow and forest floor percolate were filtered through a paper-filter (Schleicher and Schüll no. 604). The first 50 ml of the percolate were discarded. Ground water samples were filtered through a 0.2μm membrane filter (Schleicher and Schüll BA83). Soil solution samples were not filtered again in the laboratory.

To minimize changes in chemical composition of water samples were analysed as soon as possible. The fortnightly samples that had to be pooled to monthly samples later, were kept in a refrigerator at 4°C before pooling. During analysis the samples were kept at the lab-temperature, but as much as possible in the dark. Complete analysis was generally done within about one week.

After filtering the pooled samples (above-ground water and soil solutions at 0 cm depth), or within one day of sampling, pH and conductivity were measured. Next the water sample was divided in two parts. One part was used for analysis of inorganic carbon, organic plus inorganic carbon, fluoride, chloride, nitrate, sulphate, ammonium and phosphate. The other part was acidified to pH 2 or 3 with hydrochloric acid and used to determine potassium, sodium, calcium, magnesium, iron, manganese, aluminium and silica.

Short descriptions of analytical methods are given below. For details reference is made to Begheyn (1980).

pH was measured with an Orion Research pH/mV meter (model 801A) and an Orion combination pH electrode (no. 91-55), using buffer solutions of pH 4.00 and pH 7.00, made up from ampoules (Merck, Titrisol no. 9884 and 9887). A few drops of saturated potassium chloride were added to weakly buffered samples (rain water) before pH measurement.

The electrical conductivity (E_c) was measured with a Philips conductivity-meter (type PW 9501) and conductivity-cell (type PW 9513). Calibration was done with a 0.005 M KCl-solution. The conductivity is expressed as μS/cm.

Inorganic carbon was determined with a Total Organic Carbon Analyser (T.O.C. analyser. Beckman type 915 B). 50 μl of the sample was injected into the inorganic carbon channel (155°C), where carbonates and bicarbonates change into carbon dioxide under the influence of phosphoric acid. CO_2 was measured by infrared-spectrometry. Recorded peaks were compared to those of standards (0-4 mmol/1 of C). Standards were prepared from a stock solution containing 756 mg $NaHCO_3$ and 106 mg Na_2CO_3 in a volume of 1.000 litre (CO_2-free) water. The pH of this solution should be 9.2.

Total organic plus inorganic carbon was determined using the same T.O.C. analyser. The sample (50μl) was injected into the total carbon-channel (950°C) containing cobalt to catalyze the oxidation to CO_2. Carbon dioxide was measured with infra-red spectrometry, using 0-10 mmol/l C as standard solutions, prepared from a stock solution, containing 2.551 g $KHC_8H_4O_4$ in a volume of 1.000 liter (CO_2-free) water. Beckman instructions were followed.

Fluoride, chloride, nitrate and sulphate were determined by Ion-chromatography (Dionex IC 10). After injection (autosamper Cenco type 34517-040) a selective separation of anions takes place over an anion-exchange-resin (separator-column), followed by removing of cations in a fiber-suppressor column. The electrical conductivity of the successively percolating anion-acids is used to quantify the anion concentrations. The recorded peak-heights were compared to those of standards made up of stock solutions of 1.000 g/l of respectively NaF, NaCl, $NaNO_3$ and K_2SO_4 (pro analysi and dried at 105°C).

Ammonium and phosphate were determined with a dual-channel autoanalyser (Technicon AA II). After adding reagents and heating (37°C), the colour of the final solution was measured with a colorimeter. Peakheights were compared to those of standard solutions, made up of stock solutions of resp. $(NH_4)_2SO_4$ and KH_2PO_4 (both dried at 105°C). Ammonium was determined with salicylate, nitroprusside hypochlorite. Phosphate was determined with ammonium molybdate, antimony-potassium tartrate and ascorbic acid as reagents.

Aluminium and silica were also determined with a dual channel autoanalyser (Technicon AA II). Aluminium was analysed with pyrocatechol violet, ammonium-acetate/acetic acid buffer, and silica with ammonium molybdate and ascorbic acid as reagents. Standards were prepared from ampoules, containing 1.000 g (Merck Titrisol no. 9967 (Al) and no. 9947 (Si)). Technicon methods and manual instructions were followed for all autoanalyser analyses.

Calcium and magnesium were determined by atomic absorption (Perkin Elmer AAS/AES type 560) at 422.7 nm (Ca) and 285.2 nm (Mg). Lanthanum chloride was added to reduce interferences, in standards and samples (final La concentration was 1%). Standards were prepared from ampoules, containing 1.000 g (Merck Titrisol no. 9943 (Ca) and no. 9949 (Mg)). Iron and manganese were also determined by atomic absorption at 248.3 nm (Fe) and 279.5 nm (Mn). Standards were prepared from ampoules containing 1.000 g (Merck Titrisol no. 9972 (Fe) and no. 9988 (Mn)). Sodium and potassium were analysed by flame emission (PE 560). Ionization was controlled by adding aluminium solutions to standards and samples (final Al concentration).

Sodium was measured at 589.0 nm and potassium at 766.5 nm (using UV filter). Standards were prepared from ampoules containing 1.000 g (Merck Titrisol no. 9927 (Na) and no. 9924 (K)). Perkin Elmer methods and manual instructions were followed for all atomic absorption and flame emission analyses.

In some cases, the concentrations of organic anions and of bicarbonate were estimated from pH, total dissolved organic carbon and total dissolved inorganic carbon, using known (HCO_3^-) or estimated (organic anions) pK values. For organic anions, the pK was calculated using the method of Oliver et al. 1983), assuming a carboxyl content of 6 (instead of 10) μmol per mg dissolved organic C.

Concentrations of charged solutes reported usually refer to the number of moles of a unit charge of ionic substance and so are reported as equivalent ionic concentra-

Table 18. Detection limits of the water analyses, expressed in ionic equivalent concentrations.

EC	1 μS/cm	Mg	1 mmol/m³
inorganic C	3 mmol/m³	NH_4	1 mmol/m³
total C	4 mmol/m³	F	1 mmol/m³
SiO_2	1 mmol/m³	Cl	1 mmol/m³
K	1 mmol/m³	NO_3	1 mmol/m³
Na	1 mmol/m³	SO_4	1 mmol/m³
Ca	5 mmol/m³	H_2PO_4	0.1 mmol/m³

tions. They are expressed in SI units (mmol/m³ or mol/m³) but are numerically equal to the old units meq/m³ or eq/m³. Analytical detection limits are given in Table 18.

To check the analyses, six artificially made samples of known composition, similar to the natural samples were prepared according to Begheyn, 1981. Two of such standard samples were analysed during every analytical run.

The results (Table 19) show a reasonable to good recovery (95%-105%) of solutes present in relatively high concentration (> 50 mmol.m^{-3}), except for chloride in the artificial soil solution (recovery = 110%). At lower concentrations some elements are often underestimated, particularly potassium (80-90%), ammonium (80% in the artificial soil solution) and, most seriously, fluoride (40-80%) and phosphate (30-120%). The coefficients of variations are reasonable (< 10%) for most dominant solutes except chloride (8-20%), but much poorer when concentrations are below 50-100 mmol/m³. They are particularly poor for fluoride (30-200%) and phosphate (42-200%). The poor results for potassium are mainly due to analysis in 1981, and the analytical performance has increased considerably since then. Analysis of the trace elements F and P was never optimized.

4.4 Litterfall *(Th. Pape)*

At each of the four plots tree litter was sampled from funnel-shaped, sheet-plastic collectors, of 0.5 m² collecting surface area with the top about 50 cm above the soil surface. Per plot 6 collectors were placed randomly (see Fig. 19).

Sampling was fortnightly in spring and autumn, and monthly in summer and part of the winter. Samples were kept frozen in plastic bags until ovendrying (70°C), weighing and analysis. Botanical analysis of the samples was carried out by H. van Oeveren at the department of Vegetation Science, Plantecology and Weed Science of the Agricultural University, Wageningen. Chemical analysis was carried out at the department of Soil Science and Plant Nutrition at the same university according to Houba et al (1985). In 1980/81 the tree-litter was sorted into material from different trees: oak (*Quercus robur*) and birch (*Betula pendula*) in all sample plots and,

Table 19. Results of routine analysis of water standards (January 1981 to May 1984), $\bar{x}$ = mean ionic equivalent concentration ($mmol.m^{-3}$, except for EC ($S.cm^{-1}$) and pH), c.v. = coefficient of variation (%), rec = recovery of added substance (%), n = number of samples, n.a. = not applicable and – = not added to the standards.

	n = 50 throughfall			n = 26 stemflow			n = 47 rainwater			n = 22 groundwater 1			n = 23 groundwater 2			n = 10 soil solution		
	x	c.v.	rec.	x	c.v.	rec.	x	c.v.	rec.	x	c.v.	rec.	x	c.v.	rec.	x	c.v.	rec.
pH	6.35	4.3	n.a	6.61	2.9	n.a.	4.45	3.4	n.a.	6.30	5.1	n.a.	6.53	7.5	n.a.	4.25	3.3	n.a.
EC	187	4.1	n.a.	324	4.3	n.a.	52	9.8	n.a.	726	5.4	n.a.	663	4.2	n.a.	262	4.8	n.a.
Na^+	212	6.5	96.4	312	4.5	96.0	66	9.9	94.9	324	5.8	96.6	327	6.6	97.6	254	6.4	92.2
K^+	190	12.7	92.6	341	4.7	97.5	5	30.4	85	42	22.0	81.9	50	13.5	98.9	168	6.5	95.1
Ca^{2+}	125	9.5	104.1	179	6.4	102.2	43	15.8	108.3	6721	3.9	96.0	6854	3.2	97.9	504	4.3	100.9
Mg^{2+}	76	5.8	101.2	126	6.3	100.9	30	10.4	99.7	787	2.6	98.4	786	4.5	98.2	298	4.1	99.3
NH_4^+	823	8.5	96.8	1634	3.2	100.5	137	8.4	97.6	24	32.2	94.3	–	–	–	54	29.1	71.9
Al^{3+}	–	–	–	–	–	–	–	–	–	–	–	–	–	–	–	966	9.8	96.6
F^-	15	63.5	75.4	16.2	93.1	46.2	1.7	102	42.5	3.9	181	38.9	9.5	135	950	20.7	35.2	82.8
Cl^-	208	12.2	104.1	396	8.1	103.7	84	19.7	103.1	333	11.9	102.5	332	9.1	102.2	248	14.7	110.4
$0.5\ SO_4^-$	951	6.1	102.8	1689	3.2	100.9	174	6.6	102.6	1825	3.4	101.4	4356	2.9	100.1	944	6.0	104.8
NO_3^-	208	6.3	103.9	318	8.3	105.9	73	9.0	103.9	2735	4.8	107.2	–	–	–	1197	4.7	101.9
H_2PO_4	3.8	42.1	76.4	5.9	52.3	74.1	0.3	191	30.0	0.5	169	49.0	1.2	110	116	1.1	138	54.5
H_4SiO_4	–	–	–	–	–	–	–	–	–	256	6.6	102.5	249	4.7	99.7	–	–	–

additionally, poplar (*Populus tremula*, *Pop. spec.*) and alder (*Alnus glutinosa*) for plot D. Also traces of beech (*Fagus sylvatica*) and hawthorn (*Crataegus monogyna*) were found.

For the presentation of data the amounts of litter are expressed in kg per collector per plot, the contents of elements in mmol/kg and element fluxes in kmol/ha.

5 Computation of above and below ground hydrology

(J.J.M. van Grinsven)

5.1 Introduction

One of the main aims of the research project is to calculate chemical budgets. A chemical budget summarizes net fluxes for relevant chemical compounds for discrete time and depth intervals. Important chemical fluxes are fluxes of dissolved compounds in rain water, in throughfall water and in soil drainage water. For rain- and throughfall water, the solute fluxes follow directly from measured concentrations and measured water fluxes. For the soil solution monthly water fluxes were calculated by multiplying a water flux, obtained by simulation, and a measured solute concentration. Monthly soil solution samples were assumed to represent the soil solution chemistry for the preceeding and following fortnights. Monthly, integrated unsaturated soil water fluxes as a function of time and depth were calculated with a simulation model. Primary input data were precipitation and potential evapotranspiration. A deterministic model for soil water flow, based on the Darcy flow equation, was considered appropriate for the simulation to achieve the following aims:

1. to integrate fluxes over time intervals of about one month (intervals in which flux calculations are sensitive to water storage changes);
2. to realistically estimate the transpiration reduction as a result of limited soil water availability;
3. to account for capillary rise during the growing season;
4. to calibrate and validate the flow model, using measured pressure potentials.

This section describes the flow model used, the procedures used to obtain the input data, the calibration of the unknown model parameters and the validation of the calibrated values of these parameters.

5.2 The flow model

The SWATRE model (Feddes et al., 1978, Belmans et al., 1981, Belmans et al., 1983) was used for the calculation of unsaturated soil water fluxes. SWATRE was developed originally to calculate actual evapotranspiration from standard meteorological data. For this purpose unsaturated soil water flow had to be modelled to calculate reduction of transpiration due to water stress. After slight modification of the computer code, unsaturated water fluxes could be calculated.

The SWATRE model is one-dimensional, so lateral transport of water and energy (advection) and spatial variability of infiltration and soil properties were not considered. Table 20 reviews the boundary conditions and system parameters used.

5.3 Evapotranspiration

Atmospheric data necessary for the calculation of potential evapotranspiration according to conventional one-dimensional methods (Penman, Monteith-Rijtema etc.) were not measured at the study site but were taken from the weather station Winterswijk, 40 km northeast of the Hackfort location. Because local advection is probably very important in this small woodland surrounded by pasture, and because there was no opportunity to apply three-dimensional evapotranspiration models and to carry out field measurements needed to account for local advection, simple empirical equations were used to calculate potential evapotranspiration. The calculations were based on 10-day cumulative values of open water surface evaporation according to Penman.

Table 20. Boundary conditions and input parameters for the SWATRE model.

Input data	**Source**
Top of soil system	
Throughfall	Daily throughfall was calculated from cumulative daily precipitation and fortnightly throughfall, assuming an interception capacity of 0.8 mm.
Potential evapotranspiration	Calculated with empirical formulations using Penmann open water evaporation and field data of fortnightly cumulative throughfall.
Bottom of soil system	
Groundwater level	Linear interpolation of weekly field measurements.
Soil hydrology	
Water retentivity function	Optimization of closed form equation using experimental data from undisturbed soil columns. One time correction for average field hysteresis state.
Conductivity function	First estimate from water retentivity function by Mualem's method, calibrated with experimental value at −10 mbar pressure head. Fine tuning with SWATRE.
Root water uptake distribution	First estimate from root mass distribution. Fine tuning with SWATRE. Hyperbolic reduction of transpiration with pressure head below reduction point (h_3).
Initial condition	Equilibrium with phreatic level at start of growing season.

The empirical equations are as follows:

$$ET_{pot} = F_1 . E_o \quad (1)$$

$$E_{soil,\,pot} = F_2 . ET_{pot} \quad (2)$$

$$T_{pot} = (1\text{-}F_2) . ET_{pot} - (F_3 . E_{leaf,\,act}) \quad (3)$$

ET_{pot}: potential transpiration for a case without interception, soil evaporation, and advection

E_o: potential evaporation from an open water surface according to Penman

$E_{soil,\,pot}$: potential soil evaporation

$E_{leaf,\,act}$: actual evaporation of intercepted water; which is the measured difference between gross and net precipitation

F_1: 'crop' factor

F_2: soil cover parameter

F_3: ratio of the actual rate of transpiration and the actual rate of evaporation of intercepted water ($F_3 < 1$ due to advection effects).

Equations 1 and 3 account for the enhanced evaporation of intercepted water due to local advection. Transpiration is a rather conservative process which is dominated by atmospheric resistance for low vegetations, and by surface resistance for forests. When considering one-year periods, the transpiration component of evapotranspiration is fairly independent of atmospheric demand and vegetation type. For forests, transpiration roughly amounts to 350 mm; primarily depending on age and leaf area index (Roberts, 1983). Therefore advection effects on transpiration will be small in areas with a complete vegetation cover in periods without precipitation. Average daily evaporation rates of intercepted water however can exceed transpiration rates by factors as high as 3 (Singh and Szeicz, 1979).

Values of F_1, F_2 and F_3 are summarized in Table 21. F_1 applies to a situation without local advection. F_1 varies from 0.8 for a poor deciduous stand to 1.2 for a healthy coniferous stand. If F_1 exceeds 1.0 this indicates a regional advection. Values for F_1 equal to 0.7 apply to transient periods when the leaf canopy develops in spring, and disappears in autumn. Values for F_2 and F_3 are based on field observations.

Table 21. Values of evapotranspiration parameters used in equations 1-3.

time period (month. day)	F_1	F_2	F_3
01.01-05.07	0.5	1.0	0.8
05.08-05.21	0.7	0.5	0.8
05.22-09.30	0.9	0.0	0.8
10.01-10.15	0.7	0.5	0.8
10.16-12.31	0.5	1.0	0.8

Enhanced evaporation of intercepted water implies that ET_{pot} (equation 1) is smaller than 'total' potential evapotranspiration, ET^*_{pot} being the sum of T_{pot}, $E_{soil,pot}$ and $E_{leaf,act}$:

$$ET^*_{pot} = ET_{pot} + (1 - F_3) . E_{leaf,\ act}$$

A more straightforward calculation of potential evapotranspiration is difficult, because it is not possible te define 'potential evaporation of intercepted water' ($E_{leaf,pot}$) as evaporation of intercepted water is strongly connected with advection, there is no distinct upper limit to $E_{leaf,pot}$.

The nature of $E_{soil,pot}$ is rather complicated too. In practice, and also in the description of soil evaporation in the SWATRE model, soil evaporation is generally limited by the hydraulic conductivity of the surface soil and not by atmospheric boundary conditions (Black et al., 1969). Comparison of $E_{soil,pot}$ and $E_{soil,act}$ calculated by the SWATRE model showed a reduction of 50%. Theoretically, the remaining radiation energy, which is not used for soil evaporation, is available for evaporation of intercepted water.

5.4 Precipitation

For SWATRE, daily throughfall data are needed. Daily throughfall values were calculated from daily rainfall and fortnightly throughfall data assuming an interception capacity of 0.8 mm (Mitscherlich, 1971). This calculation also gives the daily interception data which are necessary for the calculation of potential transpiration. Fortnightly throughfall data from replicate collectors show little spatial variability (see Table 26), and were simply averaged.

Optimization of the interception capacity from hourly gross precipitation and fortnightly throughfall measurements, gave values ranging from 0.1 to 1.1 mm for the period from April 11 to November 3, 1981 (the throughfall-weighted mean was 0.58 mm). These values agree with the literature values, but are of uncertain quality due to (1) long throughfall collection intervals, (2) uncertainty about values for evaporation of intercepted water during the night, and (3) the presence of additional interferences (e.g. caterpillar infestations) (Arts, 1986).

5.5 Soil Hydraulic properties

Water retentivity functions ($h(\Theta)$) and conductivity functions were obtained by Van Genuchten's (1980) optimization programme, using water retention data and conductivity data from measurements on undisturbed soil cores.

Soil layers were defined for all soil plots on the basis of water retention data. Average water retention data, calculated for these newly defined layers (Table 22), were fitted by Van Genuchten's (1980) closed form equation. Integration of the closed form retentivity equation according to Mualem gives the conductivity function:

$$\theta = (\Theta - \Theta_r) / (\Theta_{sat} - \Theta_r)$$

$$h(\Theta) = (\Theta^{-1/m} - 1)^{1/n} \ . \ a^{-1}$$

$$K(\Theta) = C \ . \ \Theta^{1/2} \ . \ (1-(-\Theta^{1/m})^m)^2$$

$$C = K_{rel}(\Theta_i) / K_{exp}(\Theta_i)$$

Θ = volumetric water content ($m^3.m^{-3}$)
Θ_{sat} = saturated water content ($m^3.m^{-3}$)
Θ_r = residual water content ($m^3.m^{-3}$)
θ = relative dimensionless water content
h = pressure potential (mbar)
K = hydraulic conductivity ($m.d^{-1}$)
K_{rel} = K/K_{sat} for Θ_i
K_{exp} = experimentally determined K for Θ_i ($m.d^{-1}$)
m, n, a = parameters

Theoretical conductivity functions were fitted with the experimental conductivity at 10 mbar pressure potential (reference conductivity). It is common to use the saturated conductivity for this purpose. The advantage of using a K at low suction, is the exclusion of the contribution of macropores. Contribution of macropores to

Table 22. Average water retention data (volumetric water content in % at various pressure potentials, in mbar) for the selected soil layers.

plot and soil depth (m)	K_{10} ($m.d.^{-1}$)	pressure potential (mbar) 0	-10	-32	-100	-200	-500	-2500	-16000
A									
0.00-0.05	–	59.7	50.1	44.6	36.3	30.5	26.6	20.5	17.5
0.05-0.40	0.59	51.6	40.2	32.3	24.0	20.8	18.7	14.2	10.9
0.40-0.80	0.17	43.6	36.9	32.6	19.7	15.7	13.3	8.9	6.2
0.80-1.20	3.25	40.9	39.4	37.8	15.3	10.1	7.4	4.3	2.7
B									
0.00-0.10	–	64.7	51.2	44.2	33.2	26.2	22.4	19.2	15.7
0.10-0.50	0.68	57.4	44.1	36.1	24.3	19.5	16.3	10.8	8.1
0.50-0.80	3.21	40.6	33.5	28.0	13.2	9.8	8.3	5.7	4.1
0.80-1.20	14.00	39.5	33.7	27.3	15.3	12.4	9.2	4.6	3.1
C									
0.00-0.10	–	59.0	46.8	36.6	27.5	22.9	19.5	11.9	10.9
0.10-1.20	10.00	43.0	35.7	21.6	8.9	6.8	5.4	3.8	2.5
D									
0.00-0.20	–	58.0	47.0	35.6	23.3	19.0	15.8	13.1	10.8
0.20-1.20	0.95	42.1	36.0	29.8	14.7	9.9	7.9	5.0	3.5

conductivity is not accounted for in Mualem's approach and their presence increases the variability of the saturated conductivity considerably.

Nevertheless the theoretical conductivity function is still extremely sensitive to suction at h values around 10 mbar: conductivity can decrease by several orders of magnitude between h = 0 and h = 10 mbar. Due to uncertainties in h in the order of one millibar during the measurement of K by the two plate steady state method, errors in K(Θ) can be considerable.

Uncertainty about hysteresis in the field is an additional source of error for K. Hysteresis is not considered in the SWATRE model. Without dynamic simulation of this phenomenon, hysteresis can partly be accounted for by using water retentivity functions for the 'average field hysteresis state'. The water retention data determined in the laboratory refer to desorption of almost completely saturated soil, and do not apply to this average field hysteresis state: saturation is hardly ever attained in the field. Furthermore, the distinct air entry values usually observed when desorbing water from completely saturated cores are not found in adsorption or field experiments (Hillel, 1980). Therefore, the original laboratory desorption characteristics were used to estimate three theoretical field water retention characteristics:

1. for field desorption,
2. for field adsorption and
3. an average field water retention characteristic.

Set (1) was calculated from the original data by assuming that the saturated water content for the field was the average of total porosity and the water content at 10 mbar pressure potential. Set (2) was calculated by additionally assuming that the pressure potential associated with the water content at air entry (arbitrarily fixed at – 100 mbar) is reduced by a factor of 2 (Bloemen, 1980). The ratio between pressure

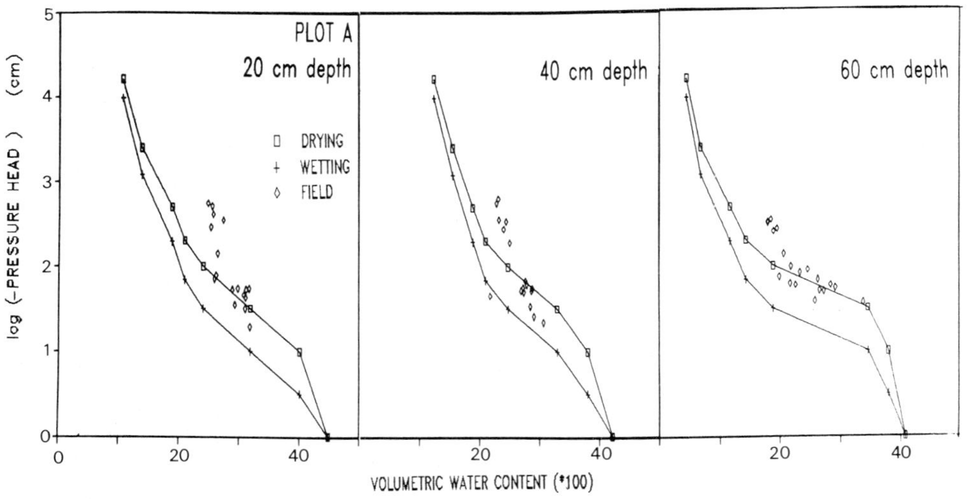

Figure 22. Comparison of water retentivity curves (h(Θ)), for plot A at 20, 40 and 60 cm depth, obtained by equilibration of 100 cm^3 undisturbed soil cores in the laboratory and obtained in the field from simultaneous measurements of volumetric water content (gamma attenuation method) and pressure head (tensiometry).

Table 23. Van Genuchten parameters for the laboratory desorption retentivity curve and, theoretically derived from this curve, the field desorption retentivity (1), the field adsorption retentivity (2) and the average field retentivity curve (3).

plot	depth (cm)	curve number	θ_{sat}	θ_f	a	N	SSQ
A	0-15	lab	0.597	0.050	0.0956	1.252	0.0043
		1	0.549	0.056	0.2803	1.180	0.0026
		2	0.549	0.047	0.0566	1.229	0.0053
		3	0.549	0.049	0.1125	1.254	0.0054
	15-40	lab	0.516	0.000	0.2903	1.216	0.0017
		1	0.459	0.001	0.6754	1.194	0.0033
		2	0.459	0.000	0.1994	1.201	0.0009
		3	0.459	0.000	0.3067	1.202	0.0015
	40-80	lab	0.436	0.019	0.0696	1.380	0.0017
		1	0.403	0.010	0.2225	1.325	0.0026
		2	0.403	0.012	0.0563	1.372	0.0014
		3	0.403	0.024	0.1214	1.369	0.0016
	80-120	lab	0.409	0.000	0.0169	2.220	0.0058
		1	0.402	0.005	0.0938	1.945	0.0111
		2	0.402	0.000	0.0340	1.638	0.0044
		3	0.402	0.017	0.0969	1.604	0.0116
B	0-10	lab	0.647	0.000	0.2145	1.212	0.0042
		1	0.580	0.007	0.4177	1.216	0.0065
		2	0.580	0.026	0.1442	1.226	0.0037
		3	0.580	0.000	0.2861	1.194	0.0041
	10-50	lab	0.547	0.007	0.1710	1.297	0.0012
		1	0.508	0.000	0.3889	1.286	0.0038
		2	0.508	0.004	0.1457	1.271	0.0007
		3	0.508	0.001	0.2155	1.270	0.0012
	50-80	lab	0.406	0.011	0.0639	1.591	0.0031
		1	0.371	0.013	0.2910	1.416	0.0047
		2	0.371	0.009	0.0550	1.581	0.0024
		3	0.371	0.030	0.1530	1.504	0.0032
	80-120	lab	0.395	0.005	0.0482	1.508	0.0013
		1	0.366	0.012	0.2419	1.400	0.0017
		2	0.366	0.003	0.0465	1.479	0.0009
		3	0.366	0.018	0.1279	1.462	0.0010
C	0-10	lab	0.590	0.151	0.2936	1.360	0.0311
		1	0.529	0.012	0.6196	1.206	0.0042
		2	0.529	0.000	0.1513	1.263	0.0031
		3	0.529	0.008	0.1702	1.291	0.0031
	10-120	lab	0.430	0.019	0.1162	1.705	0.0043
		1	0.394	0.000	0.5868	1.394	0.0105
		2	0.394	0.021	0.1619	1.590	0.0031
		3	0.394	0.015	0.2839	1.496	0.0049

Table 23 (continued)

plot	depth (cm)	curve number	θ_{sat}	θ_f	a	N	SSQ
D	0-20	lab	0.580	0.033	0.1597	1.331	0.0034
		1	0.525	0.016	0.6856	1.297	0.0184
		2	0.525	0.024	0.1652	1.290	0.0026
		3	0.525	0.009	0.2677	1.265	0.0043
	20-120	lab	0.421	0.000	0.0286	1.144	0.0117
		1	0.391	0.022	0.2483	1.470	0.0046
		2	0.391	0.000	0.0269	1.859	0.0059
		3	0.391	0.029	0.1169	1.569	0.0023

potentials of the original data set and set (2) decreased asymptotically from 2 to 1 between -100 and -10^6 mbar pressure potential. Set (3) is the arithmetic average of set 1 and set 2. All 4 data sets were fitted separately with Van Genuchten's optimization programme, giving four separate conductivity functions.

Table 23 summarizes Van Genuchten parameters for water retentivity and hydraulic conductivity functions for all soil plots and soil layers, plus the sums of squared differences (SSQ) between experimental and fitted retention data.

The theoretically derived hysteresis curves were compared with water retention data from the field. Field data were obtained from simultaneously measured water potentials (tensiometry) and water contents (gamma scintillation method). An example of such a comparison is shown in Fig. 22. The overall conclusion of this analysis was that the field desorption curve gives the best description of the average water retention behaviour in the field.

5.6 Root water uptake distribution

Root mass distribution with depth for different thickness classes was measured for all plots in 1980 (Kipp, 1981) and has been shown in Fig. 14a. As a first approach potential root water uptake (S) was assumed to be proportional to the mass distribution of roots thinner than 5 mm (Fig. 23). Roots found in the litter layer were included in the first mineral compartment.

$$S_{pot}(z) = T_{pot} \cdot (R(z)/R\,tot)$$

$S_{pot}(z)$: potential root water uptake at depth z

$R(z)$: root mass content at depth z

R_{tot}: total root mass content of soil profile

T_{pot}: potential transpiration

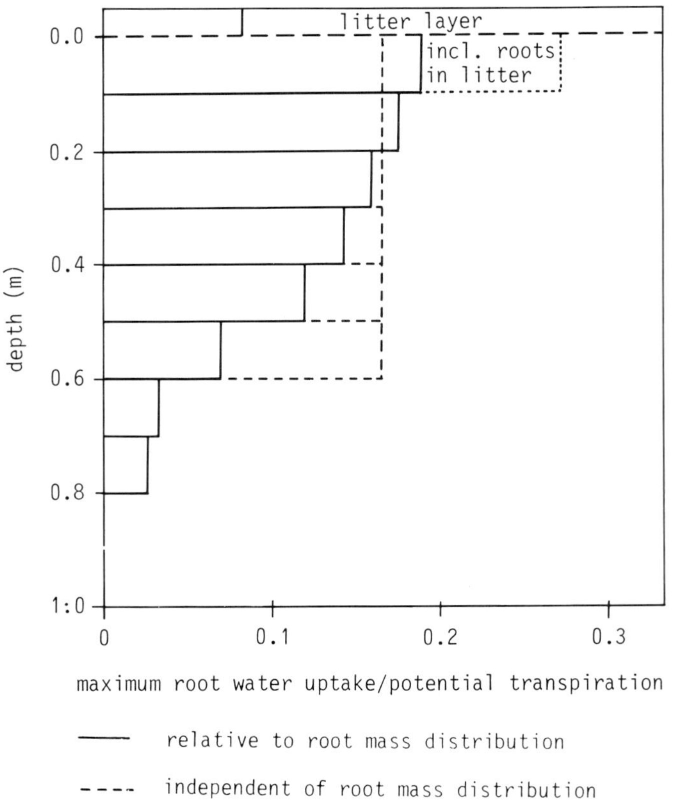

Figure 23. Average rootmass distribution with depth for plot A, B and C, and potential root water uptake distribution derived from the root mass distribution.

Actual root water uptake was calculated according to the dimensionless sinkterm variable α (Feddes et al., 1978):

$$S_{act}(z) = S_{pot}(z) \, . \, \alpha(h)$$

$S_{act}(z)$: actual root water uptake

α: dimensionless sinkterm

h: pressure potential

$\alpha(h)$ is shown in figure 24.

For $h > h_1$, $\alpha = 0$ due to aeration problems. For $h_3 < h < h_2$, $\alpha = 1$, indicating optimal water availability for the plant roots. For $h_4 < h < h_3$ transpiration is reduced due to limited water availability. The relationship between α and h is hyperbolic:

$\alpha = h_3 / h$

For $h < h_4$, $\alpha = 0$.

The sinkterm variables chosen are $h_1 = -10$, $h_2 = -25$, $h_3 = -500$, $h_4 = -16.000$ mbar.

5.7 Model calibration

Calibration of the model for a specific soil plot was carried out by optimization of the system parameters which contained the largest errors. These parameters are summarized in Table 24. Parameters F_1 and F_3 in particular will influence the absolute values of the water fluxes, while variation in conductivity and potential root water uptake distribution will mainly affect the distribution of the unsaturated water fluxes with time and depth. In the absence of direct measurements of soil water fluxes, calibration of the model towards a correct prediction of soil water flux (and thus optimization of F_1 and F_3) is impossible.

Instead of flux measurements we had a large number of field measurements of hydraulic potential as a function of time and depth for 1981 and 1982.

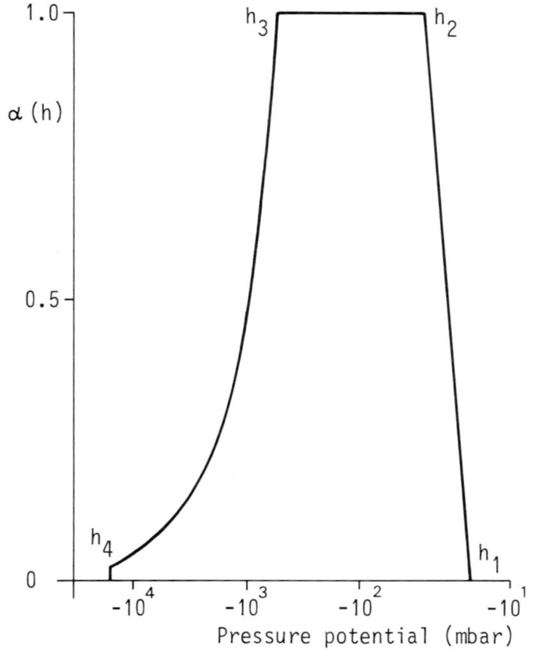

Figure 24. The relationship between the dimensionless sinkterm variable α, and pressure potential.

Table 24. Parameters calibrated with field measurements of pressure head.

Quantity	Parameter	Range of uncertainty considered
Evapotranspiration	F_1 (crop factor) F_3 (interception factor)	In the growing season: $0.8<F_1<1.2$ Average daily value: $1.0<F_3<0.33$
Hydraulic conductivity (for all soil layers)	K_{10} (reference (conductivity)	$K_{10}^{EXP} < K_{10} < K_{10}^{EXP} \bullet \frac{(K(h=0)}{K(h=-10\ mbar)}$ K_{10}^{EXP} = Experimental K (h = -10 mbar) $\frac{K(h=0)}{K(h=-10\ mbar)}$ = Theoretical ratio according to Van Genuchten model
	Van Genuchten parameters	(Effect of hysterisis considered, but not optimized by means of SWATRE)
Potential root water uptake distribution		– Proportional to root mass distribution with depth – Evenly distributed over the rootzone

There is no analytical or stochastical technique to simultaneously optimize conductivity and potential root water uptake by minimizing the difference between simulated and experimental hydraulic potential data. We tentatively calibrated the model by intuitively choosing a set of input parameters and by adjusting parameter guesses by trial and error until a satisfactory fit of the observed hydraulic potentials was obtained. This procedure could only be carried out for plot A and B. For plot C no field measurements of hydraulic potential were available. For plot D sparse hydraulic potential data were available for 1983, (so outside the calibration period).

The model calibration was first carried out for plot A. After calibrating plot A, calibration for plot B proved to be relatively simple. Hydraulic potential data for 1981 were used for calibration, those for 1982 for model validation.

An initial simulation was carried out for the period March 31, 1981 to March 22, 1982 with all parameters set at experimental or literature values. F_1 and F_3 were fixed at values given in Table 21. Potential root water uptake distribution was assumed to be proportional to the root mass distribution (Fig. 24). The reference conductivities for the Van Genuchten fits for the different soil layers were set equal to the experimental values.

The results of this initial simulation are shown in Fig. 25.1. Simulated hydraulic potentials for the upper 0.4 m of the soil are too high in periods with a large precipitation surplus, and too low when there is a large evapotranspiration surplus. Because simulation of the hydraulic potential is most sensitive to the conductivity value chosen, the SWATRE model was calibrated by changing the conductivity. The observed discrepancies between simulated and experimental pF values can be ex-

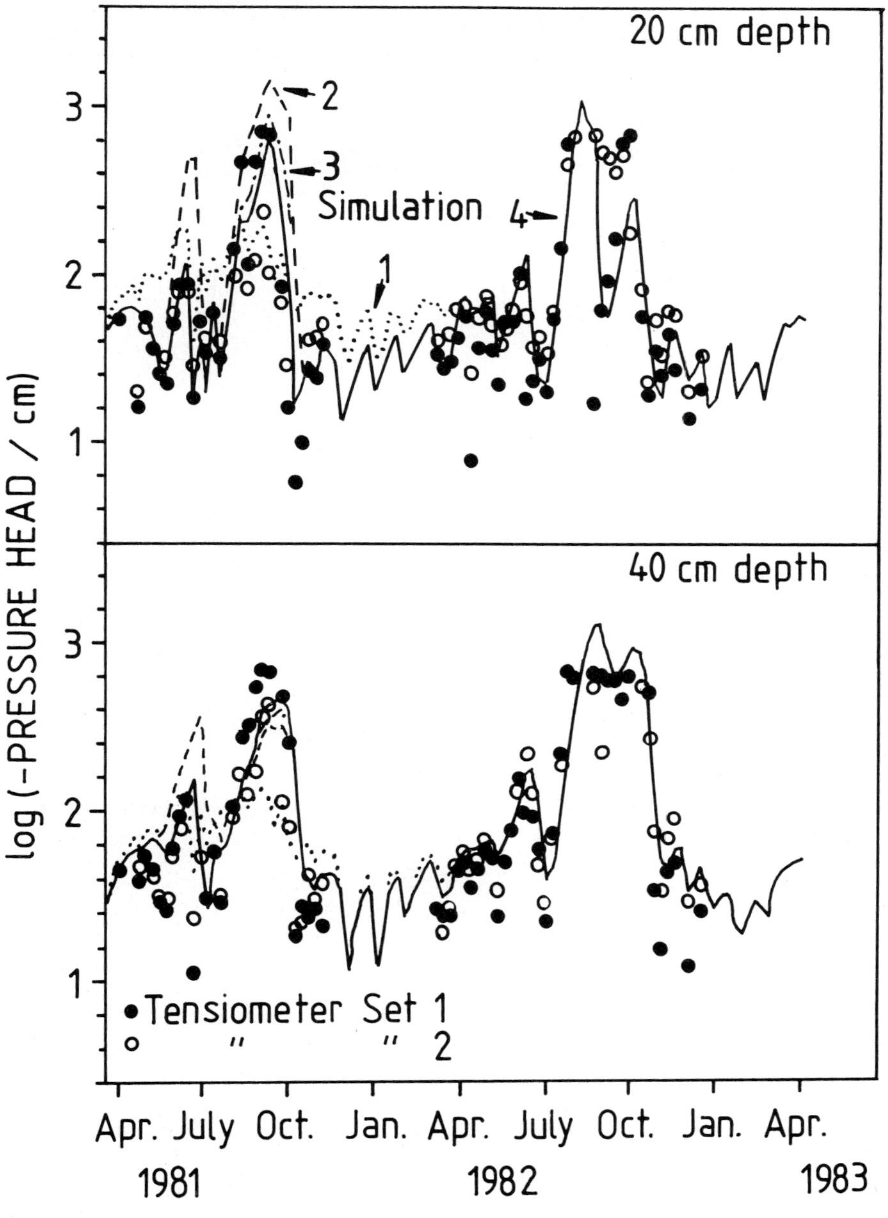

Figure 25. Pressure heads observed with duplicate tensiometers at 20 ad 40 cm depth at plot A from April 1, 1981 to March 22, 1982 and simulated by SWATRE, (1) using uncalibrated evapotranspiration parameters and hydraulic functions (.....), (2) after calibration of K(h) (----), (3) after reducing transpiration for effects of the leaf roller infestation (.__.__.__.) and (4) assuming potential root water uptake to be independent of depth (rooting depth 60 cm) instead of linearly decreasing with depth (rooting depth 80 cm) (______). The hydrological year starting at March 22, 1982 serves for model validation.

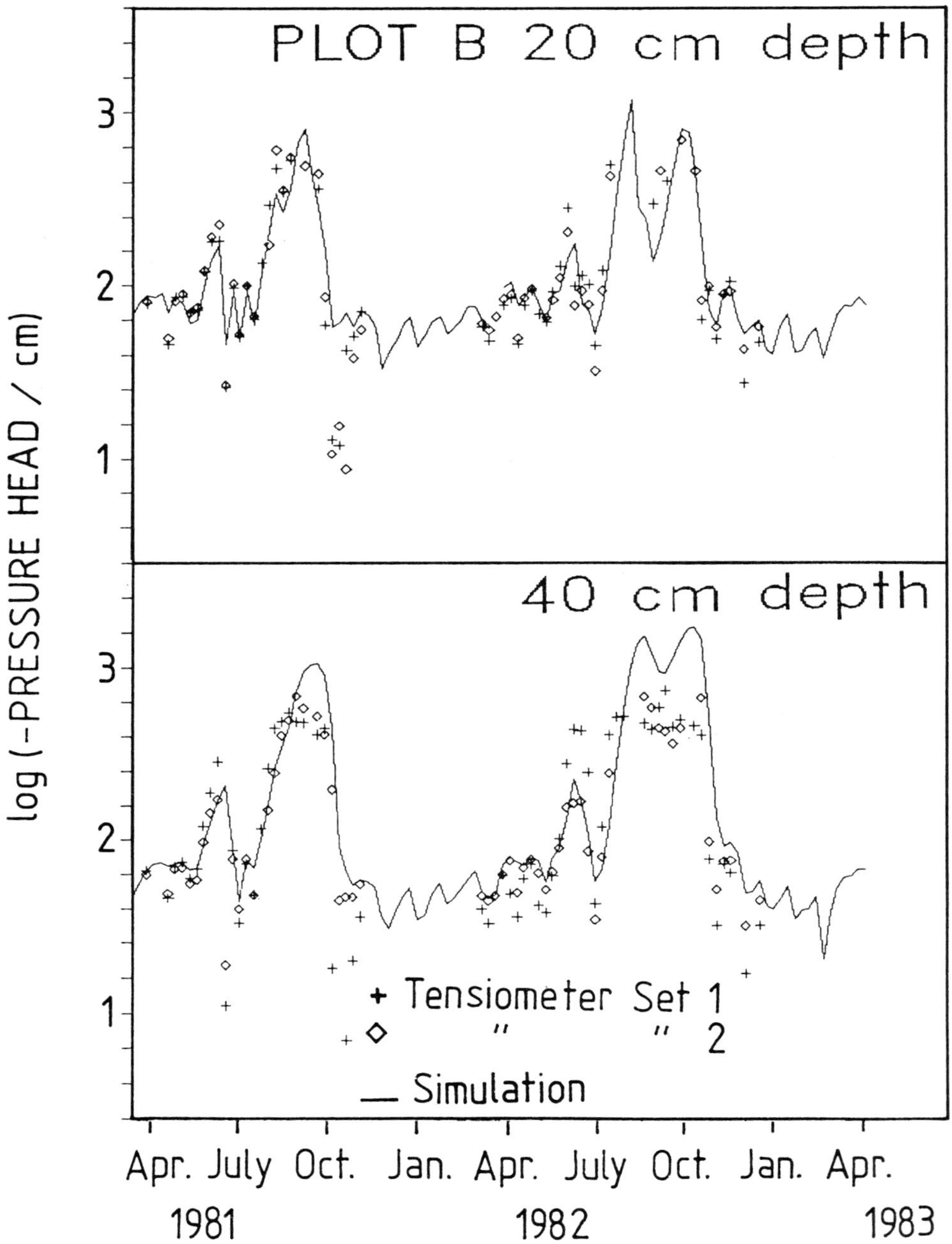

Figure 26. Agreement between observed and simulated pressure heads for plot B at 20 and 40 cm depth, using a crop factor (f1) and root water uptake distribution as calibrated for plot A, and separately calibrated K (Θ) relationships, for the period from April 1, 1981 to March 22, 1983.

plained by too high input conductivity values: in periods with high precipitation an erroneously high conductivity will lead to an overestimation of the rate of equilibration with the ground water, so that the predicted hydraulic potentials are too low. In dry periods, with excess evapotranspiration over precipitation, simulated hydraulic potentials will be too high because the rate of capillary recharge from the wet subsoil is overestimated.

Conductivity functions were calibrated by progressively decreasing the reference conductivities for all layers, until a good agreement between simulated and experimental hydraulic potentials was obtained. New sets of reference conductivities for new calibration runs were tentatively chosen using results of the preceeding runs. Using this technique a reasonable agreement between simulated and experimental hydraulic potential values was obtained (Fig. 25.2). Approximately 30 runs were used to calibrate the fitting conductivities.

The final set of fitting conductivities is not necessarily a unique solution. The simulated hydraulic potentials are especially sensitive to conductivities of the top and bottom layer of the soil, because these layers dominate the possibilities for infiltration and capillary rise. So especially with respect to conductivity of the intermediate layers calibration remains approximate. Calibrated values of reference conductivity are shown in Table 25. The discrepancy between calibrated and measured conductivity is especially large for the topsoil.

Although the evapotranspiration values for the hydrological year cannot be verified as yet, an overestimation of the evapotranspiration is obviously the reason for too low simulated hydraulic potentials from May 5 to July 7. This period coincides with infestations of the green oak (*Tortrix viridiana.*) that occurred in 1981, 1982 and 1983. These infestations resulted in severe leaf damage of oak, and caused reduced transpiration. The leaf canopy was restored near the end of June and in July. The effects of this infestation have been simulated by assuming of F_3 is 0.7 up to July 7. This considerably improved the agreement between simulated and measured hydraulic potentials (Fig. 25.3). The simulated hydraulic potentials in the topsoil, as shown in Fig. 25.3, still are slightly too low. Therefore an alternative to the root water uptake model of Fig. 23, with highest water uptake in the topsoil, was tested. Potential transpiration was assumed to be evenly distributed over the rootzone instead of proportional to root mass distribution. The alternative model slightly improved the simulation (Fig. 25.4). This result suggests that the ground vegetation, which may contribute most to the superficial root mass has a relatively low transpirative demand because of light limitation.

For plot B only the reference conductivities were optimized. Transpiration and root water uptake were quantified using the same assumptions as for plot A. Only six runs were needed to complete calibration. Results are shown in Figure 26. Hydraulic potentials at 0.6 m depth in the middle of the summer drought seem to be slightly overestimated.

Simulated hydraulic potentials in plots C and D are shown in Figs. 27 and 28 for 1981 and 1982. For plot C and D no field data were available for calibration of conductivity. Reduced values for the fitting conductivities were chosen arbitrarily, relying on experiences with the simulation for plots A and B. The conductivities for the first layer, for which no experimental data were available for all plots, were set equal to the optimized value for plot B (see Table 25).

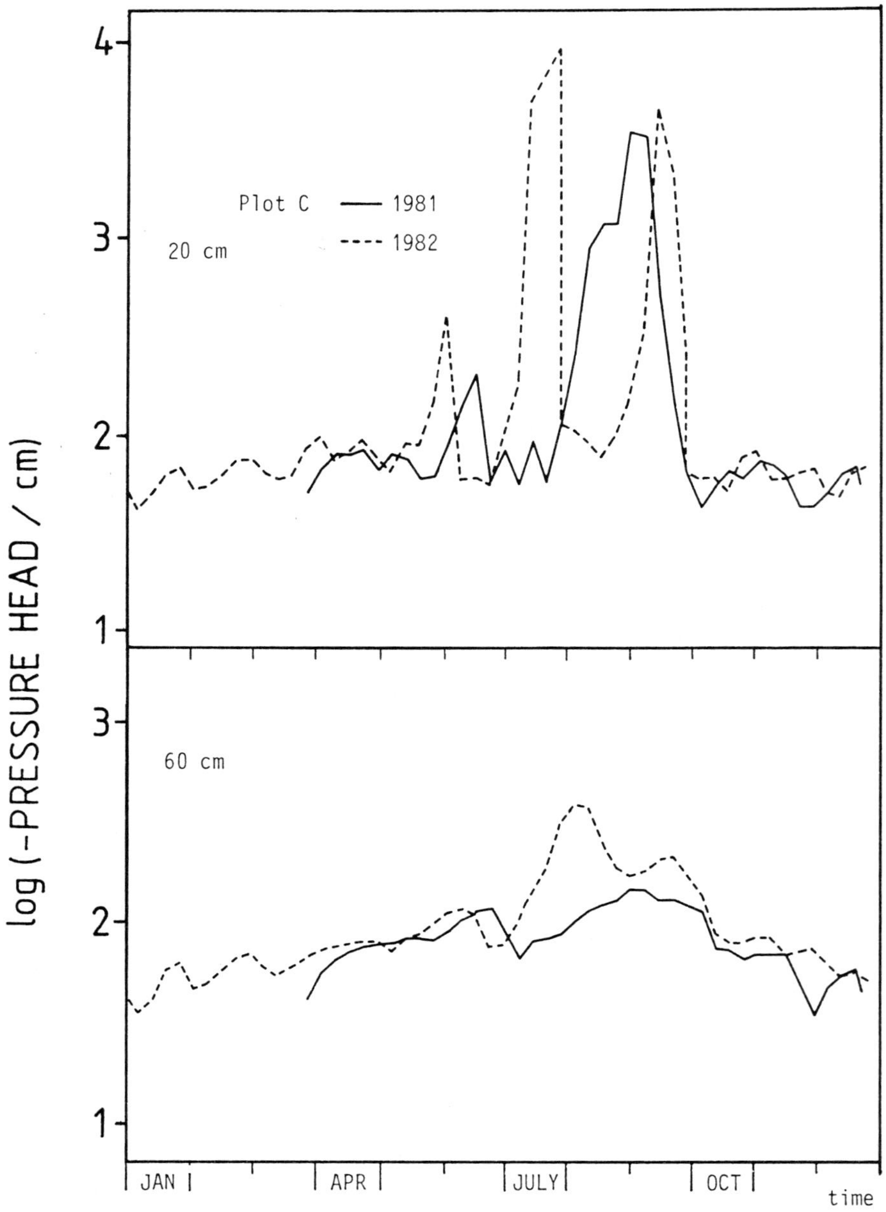

Figure 27. Simulated pressure potentials for 1981 and 1982 at 20 and 60 cm depth at plot C.

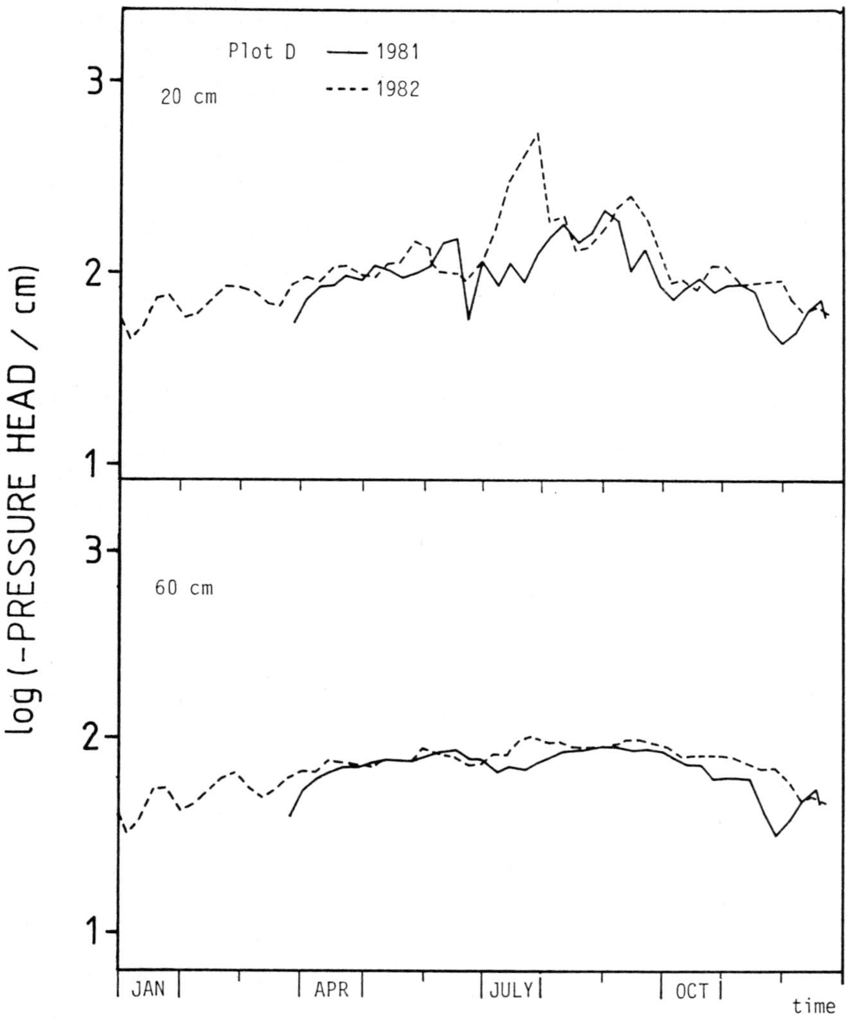

Figure 28. Simulated pressure potentials for 1981 and 1982 at 20 and 60 cm depth at plot D.

Table 25. Calibrated values for the reference conductivities.

Plot							
A		B		C		D	
depth (cm)	K_{10} (m.d^{-1})	depth (cm)	K_{10} (m.d.$^{-1}$)	depth (cm)	K_{10} (m.d.$^{-1}$)	depth (cm)	K_{10} (m.d.$^{-1}$)
0- 10	0.01	0- 10	0.10	0- 10	0.10	0- 20	0.10
10- 40	0.01	10- 50	0.10	10-160	0.20	20-160	0.25
40- 80	0.03	50- 80	0.05				
80-160	0.25	80-160	0.50				

5.8 Model validation

The period March 22, 1982 to March 21, 1983 was used for validation of conductivity, transpiration parameters and the root water uptake model, as optimized for 1981. The significance of this validation is limited because boundary conditions for 1982 and 1981, and in fact for many years are very similar. Yet a few differences between 1982 and 1981 can be pointed out. In 1982 transpiration reduction was higher than in 1981 because of limited availability of water in the rootzone, indicating insufficient capillary recharge. Furthermore the summer drought of 1982 was interrupted by a 30 mm rainstorm at the beginning of August, rewetting the topsoil.

Simulated and measured pressure potentials for 1982, for plot A are shown in Figure 25. Results for 0.2 m depth are very satisfactory. For 0.6 m depth, however, the simulated values are too high. The discrepancies could be explained by too low conductivity or too high root water uptake. However, as hysteresis and year to year variation of root water uptake distribution are not considered, no model parameters were reviewed.

Simulation results for plot B for 1982 are shown in Fig. 26. Again results at 0.2 m depth are satisfactory. However, only few hydraulic potential measurements were available for the summer drought, so that, for example, root water uptake distribution parameters could not be validated precisely. Simulation results for 0.6 m depth, however, are markedly better than for plot A, which is in line with the apparent overestimation of hydraulic potential at 0.6 m depth for plot B in 1981.

Simulation results for plot C and D for 1982 are shown in Figures 27 and 28. Site D, is markedly wetter in summer than plots A or B. Sporadic hydraulic potential measurements at plot D in 1983 confirm this. Soil profiles in plot D do have higher loam and clay content than in plot A or B which could increase capillary recharge from the subsoil.

Judging from the storage differences (which, ideally, are zero) the choice of the time boundaries of the hydrological years seems to be appropriate, except for plots C and D in 1981 (Table 37). Large storage differences at C and D for the first hydrological year result from a difference in groundwater depth; viz. 0.6 m at 1-4-1981, and about 0.8 m at 1-4-1982.

6 Results and discussion

N. van Breemen, J.J.M. van Grinsven & Th. Pape

In this chapter results of the monitoring programme for the period 1980-1985, with complete data for the three hydrological years from March 1981 to March 1984, will be presented and discussed.

6.1 Climatic and hydrological measurements

Annual precipitation and throughfall data for 1981-1984 are summarized in Table 26. For the country as a whole the years 1981, 1983 and 1984 were wetter than normal, whereas 1982 was dryer than normal. The data for the nearest station of the Royal Meteorological Service (Almen) and our own data suggest that the surroundings of the 'Oude Maat' receive somewhat less precipitation than average. Except for 1982, our precipitation values are again distinctly lower than those for Almen. The individual annual precipitation values obtained with our three collectors (the two funnel samplers for chemical characterization and the Hellman gauge) differed

Table 26. Annual precipitation (mm), measured in the open field and under the tree canopy (throughfall). Throughfall data are means ± s.d. for seven collectors per plot. For comparison, data for nearby weather stations (Almen, 5 km to the north; Deelen, 24 km to the west) and for the whole country are shown too.

year	1981	1982	1983	1984
the 'Oude Maat', open field				
a. Hellman gauge	731	659	766	730
b. Funnel # 1	727	635	802	772
c. Funnel # 2	771	633	776	696
mean for a, b and c	743	642	781	732
the 'Oude Maat', throughfall				
Plot A	598 ± 38	522 ± 33	595 ± 29	620 ± 32
Plot B	600 ± 25	514 ± 29	612 ± 47	582 ± 59
Plot C	592 ± 7	524 ± 17	605 ± 25	600 ± 36
Plot D	607 ± 39	527 ± 48	572 ± 260	642 ± 103
mean throughfall	599	522	596	611
KNMI station Almen	785	635	876	861
KNMI station Deelen	945	674	912	928
KNMI country mean	894	713	847	879

1-5% from their mean, without any collectors indicating systematically higher or lower values. Due to the so-called wind error, any precipitation gauge protruding from the land surface underestimates the true precipitation on an equivalent land surface.

For the standard rain gauge used by the Netherlands Royal Meteorological Service (KNMI), with the funnel edge at 40 cm above the land surface, the error is 3-5% (Dekker, 1979). The height of our collectors (the funnel collectors at 4 m, and the Hellman gauge at 120 cm above the land surface) probably lead to a further underestimation of the precipitation. Also (1) the proximity of a SW-NE running row of 4 m (1980) to 6 m (1985) high ash trees about 12 m southeast of the Hellmann gauge and (2) the plastic 'crowns' on the funnel collectors, may have further decreased the efficiency of the collection of precipitation. In a test at the Royal Meteorological Service, carried out at De Bilt, the Netherlands, between January and September 1983, two of our funnel collectors with plastic crowns collected only 91% of the precipitation observed by the standard rain gauge (pers. communication A.J. Frantzen, KNMI). Therefore, the precipitation data given here for the 'Oude Maat' probably underestimate the actual precipitation on the ground level in an open area by something in the order of 15%. On the other hand, the actual amount of precipitation collected by the woodland may be appreciably higher (e.g. along an edge exposed to the southwest) or lower (e.g. in the central part of the relatively narrow – 100 to 200 m – strip of forested land) than the precipitation in the open field. All of the plots except D are at least 15 m away from the edge of the woodland, while a 60 m wide strip of a rather open poplar stand is between plot D and the open field. Therefore the actual precipitation on each of the plots may be less than in the open field. In view of these uncertainties, the precipitation data for Hackfort as given will be used as such for further calculations.

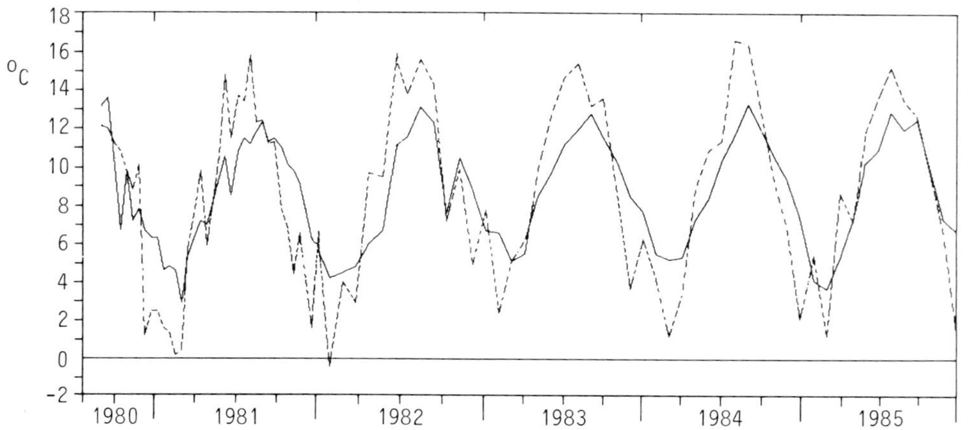

Figure 29. Soil temperatures at 5 cm depth (broken line) and 100 cm depth (continuous line), measured weekly (1980), fortnightly (1981) or monthly (1982/1985) at plot A. Temperatures measured at the other plots were generally within 0.5°C of those at plot A.

Fig. 29 shows soil temperatures at 5 and 100 cm depth, measured fortnightly or monthly from 1981 to 1985. Data from the Royal Meteorological Service show that throughout the country temperatures were rather close to normal most of the time between 1981 and 1985, except during summer 1982, which was rather dry and sunny, and summer 1983 which was exceptionally warm and dry after a very wet spring.

The groundwater table depths (Fig. 13) also reflect the peculiar weather in 1983: the drop in water table normally observed in spring started late and was particularly steep. The relatively deep watertables throughout the winter of 1984/85 may reflect the dry character of December and February during that winter.

The throughfall data are probably more reliable than those for precipitation in the open field. Because falling drops tend to be bigger and wind tends to be less in the forest stand, the wind error is probably very small. Per plot, standard deviations are generally within 5% of the average annual throughfall, and there appear to be no significant differences among the plots. The high standard deviation at plot D for 1983 and 1984 is due to anomalously high throughfall input in one collector (# 3, cf. Fig. 19) which is standing close to an overhanging tree. Throughfall is in the order of 80% of the precipitation in the open field. Throughfall is lower from spring to autumn (May 1 to October 31) than in winter (Table 27), except for plot A. That the ratio of throughfall to precipitation in plot A is similar regardless of the presence or absence of a leafy canopy may be due to a relatively low leaf area index as a result of (1) a rather open stand and (2) a relatively low vitality of the (oak) trees. The presence of quite a few birch trees and increasing decline of birch (several trees died over the measuring period) probably contributed to the open character of the stand. Over the measuring period, the canopy at A probably became more open with time: the ratio of summer throughfall at A to that at the other locations was 1.01 in 1981, 1.05 in 1982, 1.02 in 1983 and 1.10 in 1984.

The contribution of stemflow to the total precipitation reaching the ground surface in the woodland was estimated as follows. For each individual tree fitted with a stemflow trap, the relationship between the amount of stemflow (S, in liters per observation period, generally 14 days) and the precipitation collected in the Hellman gauge (R, in mm) over the same period was obtained by averaging the values of S:R observed between May 1980 and December 1981. This ratio was chosen as a measure

Table 27. Throughfall as percentage of precipitation in the open field. Values are four-year (1981-1984) means ± s.d. (n = 4) of the average throughfall for seven collectors per plot, relative to the precipitation measured by the Hellman gauge.

plot	A	B	C	D
Leaf period	80.7 ±2.6	76.7 ±4.6	77.5 ±3.0	77.6 ±4.4
Leafless period	80.7 ±5.7	83.2 ±2.6	83.5 ±4.4	91.3 ±6.0

Table 28. Precipitation reaching the ground as stemflow (mm/year) in each of the monitoring plots from April to April during the period 1980 to 1985.

	year					area considered (m^2)	number of trees
	80/81	81/82	82/83	83/84	84/85		
plot A	11.8	9.6	10.3	11.2	9.2	400	35
plot B	11.9	9.6	10.3	11.2	9.2	400	25
plot C	11.0	8.9	9.6	10.4	8.5	200	11
plot D	11.9	9.7	10.4	11.2	9.3	200	22

of the efficiency of a tree to transmit intercepted precipitation as stemflow. The period in 1980 and 1981 was chosen because later observations were increasingly troubled by problems due to leakage of the stemflow traps. The values of S:R for individual trees were between 0.01 and 0.4, and generally increased with tree size. The regression parameters of the correlation between S:R and the circumference of the tree trunk at 1.2 m height ($r=0.53$, $N=24$) were used to estimate the contribution of each individual tree to the stemflow over an area of 200 to 400 m^2 considered representative for each plot (Table 28). Despite appreciable differences in tree density, the four plots have similar stemflow fluxes with only at plot C distinctly lower values than at the other plots. Annual stemflow is always between 8 and 12 mm per year.

6.2 Solute concentrations in precipitation, throughfall, stemflow, soil solutions and groundwater

6.2.1 Precipitation, throughfall and stemflow

Table 29 gives the three-year average concentrations of solutes in precipitation, throughfall and stemflow. Data for the three nearest stations in the national network on precipitation chemistry are shown for comparison: Deelen (24 km west of 'Oude Maat'), Epe (33 km northwest) and Eibergen (25 km east). Our precipitation generally shows somewhat higher concentrations. In each type of water, the dominant ionic constituents are, in order of quantitative importance:

1. NH_4^+ and SO_4^{2-}
2. Na^+ and Cl^-
3. K^+ (only in throughfall and stemflow) and NO_3^-.

For these solutes, concentration ratios (the ratio of throughfall concentration to that in precipitation, cf. Parker, 1983) are 2 to 4. Concentrations in stemflow are 1.5-4 times higher than in throughfall. Concentration factors of 1.5-2 are typical for Na^+, Cl^- and NO_3^-, factors of about 4 are typical for NH_4^+ and SO_4^{2-}. Ca^{2+} and Mg^{2+} concentrations are generally 2 to 3 times higher in throughfall than in precipi-

Table 29. Flux-weighted means of pH, and of molar concentrations (Corg) or equivalent ionic concentrations (both in mmol/m^3) in precipitation, and in throughfall and stemflow water at the four monitoring plots. Data for Deelen, Epe and Eibergen were obtained from KNMI (Royal Meteorological Institute, De Bilt, the Netherlands), Rept 156-6 (1985).

	pH	C_{org}	K^+	Na^+	Ca^{2+}	Mg^{2+}	Fe^{2+}	Mn^{2+}	NH_4^+	Cl^-	NO_3^-	SO_4^{2-}	$H_2PO_4^-$	$\Sigma+$	$\Sigma-$
Preciptitation 'Oude Maat'															
	4.56	204	7.1	124	55	30	2.5	0.85	168	127	76	185	1.7	415	390
Precipitation Deelen (1), Epe (2) and Eibergen (3)															
1.	4.55	–	4.1	73	26	18	–	–	129	86	58	136	0.6	278	281
2.	4.56	–	4.5	69	22	16	–	–	66	78	41	82	0.2	205	201
3.	5.46	–	6.9	62	32	18	–	–	187	71	59	140	2.1	309	302*
Throughfall 'Oude Maat' (April 1981-1984)															
A	4.80	1474	197	205	128	87	3.6	12.0	474	260	172	575	12.7	1122	1020
B	4.69	1560	209	195	117	84	3.0	6.1	542	254	185	542	16.3	1177	997
C	4.62	1219	159	172	108	81	2.8	8.1	421	228	153	433	15.6	977	830
D	4.98	1873	210	261	235	124	3.1	8.8	609	381	180	705	13.5	1461	1280
Stemflow 'Oude Maat' (1981-1983)															
A	5.24	2802	268	439	206	166	6.0	15.8	2035	583	290	2125	2.9	3142	3001
B	5.18	3336	335	383	183	134	4.9	7.5	2215	554	220	2193	4.7	3269	2972
C	5.33	2586	284	403	176	140	4.0	10.3	1989	541	225	2013	4.0	3010	2783
D	5.31	2650	204	429	296	152	3.7	7.9	1803	597	208	1917	1.4	2900	2723

* incl. 30 mmol/m^3 of HCO_3^-.

tation while stemflow shows 1.5-2 times higher levels of these cations than in throughfall. Much higher concentration factors are observed for dissolved organic carbon (6 to 9), K^+ (20 to 30), Mn^{2+} (7 to 14) and $H_2PO_4^-$ (7 to 10). For these solutes a further increase in concentration from throughfall to stemflow is small (2 for organic carbon) or absent (K^+ and Mn^{2+}), whereas $H_2PO_4^-$ has distinctly lower (1/4 to 1/10) concentrations in stemflow than in throughfall. pH is slightly higher in throughfall (4.6-5.0) than in precipitation (4.6), and the highest in stemflow (5.2-5.3). Iron is the only solute considered that shows no distinct differences in concentration among the three types of water.

Throughfall compositions clearly differ among the four plots, with low concentrations for most solutes at C, and increasingly higher concentrations at B, A, and D, in that order. In 1981, sufficient replicated samples were taken to test the statistical significance of differences in (flux-weighted) mean throughfall concentrations among the four plots. Variation coefficients for the concentrations of bulk solutes were generally well below 10% at plots A and C, between 5% and 17% at B, and between 10% and 30% at D (Table 30).

The large spatial variation in throughfall concentrations at D for most solutes are due mainly to results for one pair of throughfall collectors (D3 + D4, see Fig. 19) one of which (D3) was placed close to an overhanging tree, possibly causing 'leakage' of water into the throughfall collector.

Table 30. Variation coefficients (standard deviation as ‰ of the mean of three values) for the flux-weighted annual concentrations of solutes in throughfall water at each of the monitoring plots in 1981.

plot	H^+	C_{org}	K^+	Na^+	Ca^{2+}	Mg^{2+}	Fe^{2+}	Mn^{2+}	$NH4^+$	Cl^-	$NO3^-$	$SO4^{2-}$	$H_2PO_4^-$
A	24	8	4	4	7	4	15	7	3	4	3	3	14
B	14	23	11	6	10	9	25	29	17	9	4	13	15
C	6	7	9	2	3	3	11	4	9	3	1	8	3
D	33	18	25	14	16	20	16	46	31	13	12	26	40

Table 31. Linear correlation coefficients among solute concentrations in throughfall at all plots, 1981-1984. In total, 432 samples were considered, 252 in 1981, 84 in 1982 and 48 in 1983 and 1984.

Corg	K^+	Na^+	Ca^{2+}	Mg^{2+}	Fe^{2+}	Mn^{2+}	NH_4^+	Cl^-	NO_3^-	SO_4^{2-}	$H_2PO_4^-$	
0.18	0.05	0.07	-0.01	0.06	0.14	-0.11	0.30	-0.04	-0.26	0.08	0.16	pH
xxxx	0.45	0.05	0.38	0.42	0.58	0.38	0.48	0.09	0.06	0.12	0.50	Corg
	xxxx	0.12	0.67	0.64	0.31	0.53	0.19	0.29	0.35	0.17	0.46	K^+
		xxxx	0.30	0.47	0.04	0.29	0.37	0.87	0.04	0.49	-0.03	Na^+
			xxxx	0.83	0.26	0.52	0.35	0.42	0.49	0.41	0.42	Ca^{2+}
				xxxx	0.33	0.70	0.57	0.48	0.55	0.59	0.53	Mg^{2+}
					xxxx	0.28	0.42	-0.02	0.14	0.13	0.40	Fe^{2+}
						xxxx	0.42	0.36	0.38	0.50	0.33	Mn^{2+}
							xxxx	0.22	0.20	0.77	0.47	$NH4^+$
								xxxx	0.06	0.39	-0.03	Cl^-
									xxxx	0.11	0.63	$NO3^-$
										xxxx	0.02	$SO4^{2-}$

Table 32. Linear correlation coefficients among solute concentrations in stemflow at all plots, 1981-1983. In total, 353 samples were considered, 203 in 1981, 102 in 1982 and 48 in 1983.

C_{org}	K^+	Na^+	Ca^{2+}	Mg^{2+}	Fe^{2+}	Mn^{2+}	NH_4^+	Cl^-	NO_3^-	SO_4^{2-}	$H_2PO_4^-$	
-0.03	-0.18	0.09	-0.10	-0.09	0.03	-0.14	-0.00	0.11	-0.14	-0.11	-0.03	pH
xxx	0.66	0.37	0.53	0.59	0.56	0.46	0.75	0.28	0.61	0.56	0.58	Corg
	xxx	0.46	0.61	0.70	0.49	0.59	0.69	0.37	0.60	0.66	0.63	K^+
		xxx	0.60	0.70	0.24	0.59	0.70	0.87	0.27	0.70	0.19	Na^+
			xxx	0.89	0.30	0.71	0.71	0.46	0.40	0.79	0.32	Ca^{2+}
				xxx	0.36	0.87	0.84	0.53	0.39	0.91	0.38	Mg^{2+}
					xxx	0.26	0.50	0.12	0.60	0.29	0.62	Fe^{2+}
						xxx	0.69	0.44	0.29	0.80	0.29	Mn^{2+}
							xxx	0.57	0.51	0.89	0.50	$NH4^+$
								xxx	0.14	0.57	0.07	Cl^-
									xxx	0.31	0.73	$NO3^-$
										xxx	0.28	$SO4^{2-}$

At C, pH and the concentrations of organic C, K^+, Na^+, Ca^{2+}, Mg^{2+}, NH_4^+, Cl^- and SO_4^{2-} were significantly ($P < 0.05\%$) lower than at A. At A, Mn^{2+} concentrations were significantly higher than at C and B, while D showed significantly higher Ca^{2+} and Cl^- concentrations than the three other plots. Finally, concentrations of organic carbon at D, and of NO_3^- at B, were significantly higher than at C. Other differences were not significant.

Tables 31 and 32 show the correlations between concentrations of individual solutes in throughfall and stemflow water. In both types of water, solute concentrations are similarly correlated. pH is not correlated with the concentration of any solute. The concentrations of most bulk solutes are positively correlated, with highest correlation coefficients for the pairs NH_4^+ & SO_4^{2-}, Mg^{2+} & SO_4^{2-}, Mg^{2+} & Ca^{2+}, Na^+ & Cl^- and Mg^{2+} & NH_4^+ (> 0.8). NH_4^+ and SO_4^{2-} are not only highly correlated, but throughout the (ionic equivalent) concentration range of 0.05 – 10 mmol/l in precipitation, throughfall and stemflow, their concentrations in individual samples are generally quite similar (Fig. 30).

K^+ concentrations are generally low (< 0.1 mmol/l) in winter, and increase in April to maximum values (0.2 – 1 mmol/l) between September and mid-November (Fig. 31). The concentrations of most other solutes in throughfall water also differ between winter and the rest of the year (Table 33). In addition to K^+, organic C, Fe^{2+}, Mn^{2+}, NO_3^- and $H_2PO_4^-$ are significantly higher during the vegetation period. The same is true for Ca^{2+} if the strong plot-to-plot variation (due to very high Ca^{2+} levels at D) is accounted for by standardizing the values to 100% for each plot mean. On the other hand, Na^+, NH_4^+ and SO_4^{2-} show significantly higher concentrations in winter.

Evaporation of intercepted water, dry deposition of atmospheric gases and particles, leaching of substances from above-ground parts of the vegetation, and assimilation of nutrient elements by tree leaves and by biota in the phyllosphere and on the tree bark all influence solute concentrations in throughfall and stemflow (Parker,

Table 33. Volume-weighted equivalent ionic (or, for Corg, molar) concentrations in throughfall during the vegetation period (15 April to 1 December) and the leafless period (1 December to 15 April). Means and standard deviation for all plots, January 1981 to December 1984 ($n = 16$). Units are $mmol/m^3$. Values that are significantly higher ($P < 0.05$) in a particular period than in the rest of the year have been underscored.

	H	C_{org}	K	Na	Ca	Mg	Fe	Mn	NH_4	Cl	NO_3	SO_3	H_2PO_4
Vegetation period													
mean	16	1770	255	145	151	89	4	9	421	212	213	434	19
s.d.	11	615	93	41	77	32	1	3	110	75	70	118	8
Leafless period													
mean	10	687	57	228	104	73	1	5	602	282	99	639	4
s.d.	6	247	17	104	49	30	1	3	212	158	24	234	2

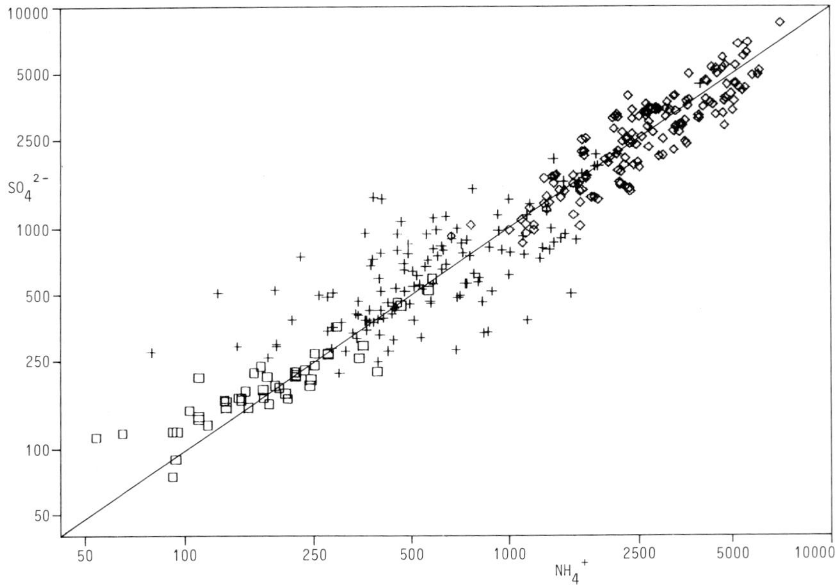

Figure 30. Ionic equivalent concentrations of NH_4 and SO_4 in precipitation (squares), throughfall (crosses) and stemflow (diamonds) in 1982 and 1983 (mmol/m^3, logarithmic scale). The line indicates equal concentrations of NH_4 and SO_4.

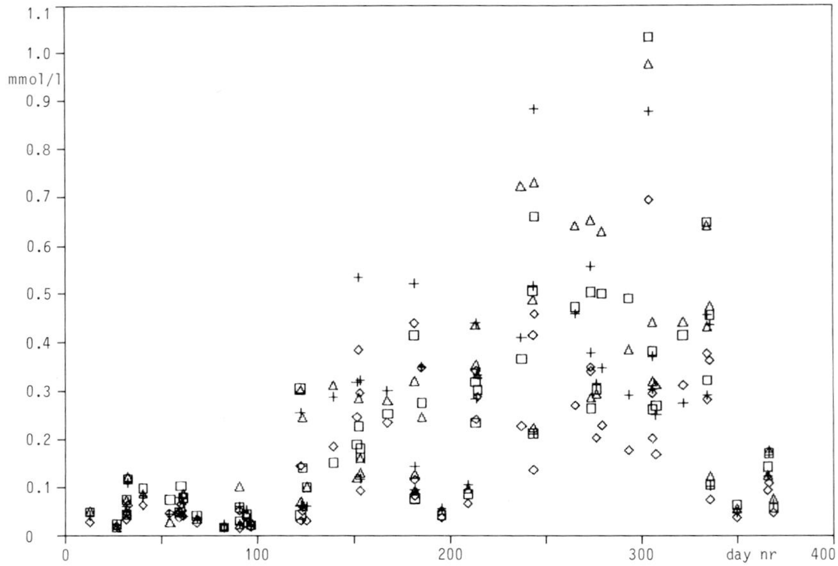

Figure 31. Concentrations of K^+ in throughfall water as a function of time of the year (day nr.) in 1981, 1982, 1983 and 1984. Squares: plot A, crosses: plot B, diamonds: plot C, triangels: plot D.

1983). These aspects will be dealt with in paragraph 6.3 where the element budgets are discussed.

6.2.2 Concentrations in the soil solution

Table 34 gives the flux-weighted mean equivalent concentrations of all major solutes at the various sampling depths between April 1981 and 1984. The variations of solute concentrations with time and depth are illustrated by the contour plots of Appendix 5. The variability in solute concentrations at a given depth is indicated by relative standard deviations (variation coefficients) shown in Table 35. The data base of Tables 34 and 35 refers to a limited number of soil solution collectors (for each depth considered generally one or two, sometimes four) per plot. Therefore, the variation indicated in Table 35 is mainly of a temporal nature. Postma (1986) studied the spatial variation in solute concentrations over the distance of the plot size, and the temporal variation within the (monthly) sampling interval. For Al^{3+}, NO_3^- and Cl^- the (spatial) variation coefficient was close to 30% at 20 cm depth, and between 10 and 20% at 90 cm depth. Sampling once per month instead of every two to four days resulted in an error in the mean monthly concentration of about 10% for Al^{3+} and NO_3^- at 50 cm depth in plot B. The data on the soil solution composition and its variation with time and depth will be discussed below for each of the various solutes separately. Processes involved in determining the soil solution composition will be dealt with briefly in paragraph 6.4.

Organic carbon is the dominant solute in the surface 20 to 40 cm of the soils at all plots. The dominant anions are nitrate and sulfate in all soils, the dominant cations are calcium (in calcareous soil horizons) and aluminum (in acidic soil horizons). In calcareous horizons, bicarbonate is an important anion. The 0 cm depth solutions (which represent water percolating from *unrooted* forest floor) differ strongly from the solutions from mineral soil horizons in having relatively low concentrations of silica and aluminum, and relatively high concentrations of ammonium and phosphorus. Variability is small for pH (CV $<$ 10%), generally moderate (CV 20 – 80%) for most solutes in concentrations above 0.1 mmol/l, and high (CV 100 – 300%) for solutes in low concentrations ($<$ 0.1 mmol/l). For the solutes that form the bulk of the dissolved matter, the analytical precision is relatively good, and most of the variability at a given depth is temporal. For the solutes in trace concentrations, analytical errors may contribute significantly to the reported coefficients of variation. The concentrations of individual solutes in soil solutions of the four plots, and variations in concentrations with time and depth will be discussed below.

Electrical conductivity

Total ionic concentration, as indicated by the electrical conductity is highest in the soil solutions of D (0.4 – 1 mS/cm) lowest in C (0.1 – 0.3 mS/cm), and intermediate in A (0.3 – 1.0 mS/cm) and B (0.25 – 0.5 mS/cm). The electrical conductivity shows a very distinct seasonal variation, with highest values in summer (surface soil) and

Table 34. pH dissolved organic C ($mmol/m^3$), and ionic equivalent concentrations in soil solutions from various depths at each of the four plots ($mmol/m^3$). Values are means of all available data between April 1981 and April 1984 (10-90 cm depth), or between February 1983 and April 1984 (0 cm depth). The number of samples involved is 30 to 40 for the 20- and 60 cm depth, about 60 for the 10- and 40 cm depth, and about 90 for the 90 cm depths.

dpth	pH	C_{org}	SiO_2	K^+	Na^+	Ca^{2+}	Mg^{2+}	Al^{3+}	Fe^{2+}	Mn^{2+}	NH_4^+	Cl^-	NO_3^-	SO_4^{2-}	$H_2PO_4^-$	HCO_3^-	A*
Plot A																	
0	3.60	3370	86	371	245	480	235	77	19	66	384	305	996	709	31.0	–	108
10	3.59	4793	417	293	248	778	311	592	34	75	239	322	1441	846	3.79	–	153
20	3.55	3408	506	228	255	572	335	985	17	64	71	297	1667	826	1.11	–	107
40	3.81	2847	580	113	353	621	353	2063	4	80	50	430	1902	1353	1.79	–	101
60	4.60	1860	695	30	491	2387	888	587	2	118	27	776	2067	1504	4.59	–	91
90	7.05	1968	422	10	494	4786	1506	61	2	2	36	572	2369	1441	1.45	2322	142
Plot B																	
0	4.20	6655	100	401	210	314	165	29	31	16	1096	310	870	695	65.4	–	280
10	3.50	5537	323	310	229	407	195	570	31	20	185	237	1231	699	4.74	–	169
20	3.51	4384	380	326	216	549	243	946	32	22	151	253	1623	735	5.73	–	135
40	3.95	2943	314	208	239	486	250	1589	5	20	105	349	1492	934	1.57	–	111
60	3.98	3682	323	218	257	508	234	1902	6	17	80	345	1685	993	0.47	–	141
90	3.99	3068	401	120	363	465	249	2144	4	14	42	539	1373	1533	0.43	–	118
Plot C																	
0	3.84	11793	191	425	196	354	224	58	47	30	974	236	1178	564	55.1	–	424
10	3.85	4975	424	223	213	294	201	585	23	21	161	259	717	719	2.7	–	180
20	3.82	4031	425	218	199	201	201	475	24	18	160	230	583	585	1.81	–	144
40	4.04	3983	482	159	225	281	368	704	10	21	91	276	769	681	0.95	–	157
60	4.12	1862	429	127	224	331	366	753	2	42	32	273	768	850	0.89	–	76
90	4.14	2282	481	67	323	344	484	1054	2	39	28	471	670	1174	0.46	–	94
Plot D																	
0	6.55	6967	144	289	274	1166	298	9	5	8	176	330	386	491	3.70	403	481
10	7.51	3647	155	161	274	4657	377	15	3	2	44	397	2238	1014	0.75	1709	263
20	7.43	3584	210	56	360	5245	442	13	4	2	34	458	2586	1222	0.77	1860	258
40	7.54	2681	176	16	311	5941	409	10	3	2	34	420	2916	1241	0.62	2167	193
60	7.18	2095	163	12	364	5251	706	6	2	1	35	482	2967	1384	0.54	1666	151
90	7.33	2069	121	10	314	6323	531	8	2	1	22	545	2703	1414	0.56	2570	149

* A = organic anions

Table 35. Coefficient of variation (standard deviation as percentage of the mean) of the concentrations of soil solutes at each sampling depth in the four plots. The number of samples involved is generally 30 to 35 for 0, 20 and 60 cm depth, 55 to 65 for 10 and 40 cm depth, and 80 to 90 for 90 cm depth.

depth	pH	C_{org}	SiO_2	K^+	Na^+	Ca^{2+}	Mg^{2+}	Al^{3+}	Fe^{2+}	Mn^{2+}	NH_4^+	Cl^-	NO_3^-	SO_4^{2-}	$H_2PO_4^-$
Plot A															
0	5	24	88	49	49	53	40	77	53	56	69	63	57	47	75
10	13	30	47	53	23	79	35	53	54	39	111	55	46	43	235
20	4	35	41	22	35	30	89	29	106	53	107	62	36	33	378
40	6	80	25	88	43	32	35	34	125	45	220	35	56	18	635
60	5	31	13	97	26	35	34	82	150	40	230	41	54	22	473
90	8	66	23	150	27	27	25	551	100	200	192	39	40	22	489
Plot B															
0	16	93	98	63	45	64	56	93	177	81	119	62	80	53	104
10	16	84	69	37	40	60	48	92	58	70	101	49	62	52	136
20	12	20	53	28	33	39	52	30	38	59	155	53	41	34	113
40	6	42	32	44	33	30	36	26	60	55	201	42	45	31	201
60	6	48	25	19	41	24	26	22	50	53	141	50	39	25	130
90	7	67	35	106	56	56	39	51	75	64	183	100	46	45	1575
Plot C															
0	9	146	64	37	39	57	59	84	96	60	111	53	67	43	76
10	10	37	54	53	45	40	40	46	83	48	98	55	60	42	102
20	12	25	64	53	31	40	42	38	58	44	106	57	67	36	194
40	9	26	34	59	44	56	52	36	40	76	191	58	87	36	307
60	6	49	23	35	26	28	21	25	100	50	228	38	51	19	358
90	6	52	27	76	37	41	35	53	100	77	243	52	63	42	480
Plot D															
0	7	92	78	156	38	27	23	144	60	88	97	54	73	46	124
10	4	41	34	58	32	40	42	140	67	150	168	68	73	50	281
20	3	53	22	48	30	31	32	146	225	200	200	66	52	48	300
40	3	42	31	69	22	25	30	170	300	200	203	54	48	40	316
60	3	40	24	83	16	28	32	183	150	200	194	44	52	32	357
90	3	49	32	130	25	17	35	238	100	200	259	36	33	28	311

fall (at greater depths) in case of soils A, B and D. At soil C, on the other hand, the lowest EC values in the surface soil occur in summer. As will be discussed later, these differences are related mainly to the differences in the dynamics of nitrate (the dominant ionic substance in most soil solution samples) between C and the three other soils.

pH

In the acid soils A, B and C, the pH increases with depth. The increase is slight in B (from 3.5 to 4) and C (from 3.9 to 4.1), and very pronounced in A (from 3.6 to 6.8) on account of its calcareous subsoil. Relatively high pH values were observed at all depths during the first one or two sampling times, which may have been because the porous cups were not yet 'conditioned'. Data in the first two samples were incorporated in the contourplots (Appendix 5), but not in Tables 34 and 35. In general, variations in pH. The pH with time are somewhat larger at 10 and 20 cm depth (CV 10 – 16%, at A, B and C) than deeper in the soil. There are no pronounced seasonal variations in pH. The pH dropped strongly during the second half of 1981 at all depths, a phenomenon that was attributed to nitrate formation (Van Breemen and Jordens, 1983), but much less so in the following years. The fact that in the calcareous soil D the pH also decreased in the second half of 1981 (from 7.5 to just below 7) makes one wonder whether a systematic analytical error was involved.

Organic carbon

Organic carbon is the dominant solute in all soils, except in D, where calcium predominates. Organic C concentrations decrease from about 5 mmol/l at 10 cm depth to about 2 (A, C, D) or 3 (B) mmol/l at depths below 60 cm. The contour plots illustrate that organic C concentrations tend to be highest in summer and fall at all sampling depths.

Dissolved silica

Silica concentrations are highest in soil A (0.5 mmol/l), lowest in soil D (0.15 mmol/l) and intermediate in B and C (0.3 – 0.4 mmol/l). They are strongly seasonal, particularly in the surface horizons, with highs in summer and fall, and lows during winter.

Potassium

In all four soils, the highest potassium concentrations are usually observed in the surface soil (0.2 – 0.6 mmol/l). The potassium concentrations generally decrease with depth, and the decrease seems to be stronger as the pH increases more steeply down the soil profile: the lowest K^+ concentrations (0 – 0.2 mmol/l) occur in the calcareous subsoils of soils A and D. In the surface soil at C and D, dissolved K^+ varies seasonally, with lowest concentrations in summer. This pattern is probably related to strong plant uptake of K^+ during the growing season, and contrasts with that for EC and most other solutes, which tend to be concentrated by evapotranspiration in the same period.

Sodium

The mean Na^+ concentrations are lowest in soil C (about 0.2 mmol/l) and progressively higher in B (0.25 mmol/l), D (0.3 mmol/l) and A (0.35 mmol/l). In soils A, B and C actual concentrations have general tendency to increase with depth, and show a distinct seasonality, reflected by high concentrations below the root zone in the second part of the year. In soil D the spatial and temporal variations are smallest. The similarity of (1) the patterns of the contour plots for Na^+ and Cl^-, (2) their concentrations and (3) the clear correlation between their concentrations ($r = 0.63$ at A, 0.73 at B and 0.88 at C, for about 290 samples at each plot), indicates that NaCl may be the source for both ions and suggests that their mobility is rather similar. The high concentrations of Na^+ relative to those of Cl^- in 1981, may have been due to contamination from the ceramic cups used for sampling soil solutions.

Calcium

On account of the simultaneous presence of calcareous soils or subsoils, and of strongly acidic leached soils in the study area, wide variations in dissolved calcium concentrations occur among the four sites. Mean ionic equivalent Ca^{2+} concentrations vary from 5 to 6 mmol/l in calcareous soil D to 0.3 mmol/l in soil C. The acidic surface soil of A (0.9 mmol/l) and soil B (0.5 mmol/l) are distinctly higher in dissolved Ca^{2+} than soil C. Ca^{2+} concentrations in B and C vary seasonally in a manner very similar to the variations in dissolved NO_3^-. The same is true for soil D, although this is not shown by the contour plot for calcium. Coefficients of correlation between Ca^{2+} and NO_3^- concentrations are: $r = 0.51$ at plot A, 0.69 at B, 0.41 at C and 0.84 at D ($n = 290$).

Magnesium

Mg^{2+} concentrations vary in a manner similar to those of Ca^{2+} (see Appendix 5), but the relative differences among the soils are smaller: highest mean (ionic equivalent) concentrations were observed in the subsoils of A (1.5 mmol/l) and D (0.6 mmol/l), followed by the surface soils of D (0.4 mmol/l) and A (0.3 mmol/l). At C (0.3 mmol/l) and B (0.22 mmol/), the mean Mg^{2+} concentrations increase only slightly with depth. Note that Mg^{2+} concentrations are higher in soil C than B, while the reverse holds for Ca^{2+}.

Aluminium

Dissolved aluminium is mainly present as trivalent aquo aluminium, and normally less than 10% is organically complexed (Mulder et al, submitted for publ). Mean ionic equivalent Al^{3+} concentrations are highest in the most acidic subsoils: about 2 mmol/l at A (40 cm depth) and B (60 – 90 cm depth), distinctly lower in the moderately acid subsoil of plot C (1 mmol/l), and virtually nill in calcareous soil D. In all acidic soils aluminium is the dominant dissolved cation on an equivalent basis. Al^{3+} concentrations increase with depth, and, at least in the subsoil, show a seasonal variation with highest values in fall and winter. Maximum ionic equivalent concentrations may be as high as 3.5 – 4.5 mmol/l. The pH also increases with depth, so there is a positive correlation between pH and the aluminium concentration in such conditions. This is somewhat unexpected because the aluminium concentration in equilibrium with various Al-bearing minerals decreases with increasing pH. As is pointed out elsewhere (Mulder et al., subm. for publ.), surface soil solutions are far undersaturated with most Al-bearing minerals (so that aqueous Al concentrations are related simply to pH). In most soil horizons dissolved nitrate is strongly correlated with aqueous aluminium (van Breemen and Jordens, 1983).

Iron

The form of dissolved iron is not known. It has been represented here as aqueous Fe^{2+}, but most of it may in fact be organically complexed FeIII. Dissolved iron is generally very low (ionic equivalent Fe^{2+} concentrations below 5 μmol/l) in subsoils and in calcareous soil, but reaches distinctly higher values (20 – 50 μmol/l) in the surface horizons of the acidic soils. As with other dissolved trace solutes (notably phosphate) the concentrations sometimes apparently vary haphazardly with time and depth, which may be due in part to analytical errors or sample contamination.

Manganese

Manganese is presumably present mainly as Mn^{2+}. Mean ionic equivalent Mn concentrations are very low (< 5 μmol/l) in calcareous soils (soil D and soil A at 90 cm depth), higher in soils B and C (15 – 40 μmol/l) and highest in the upper 60 cm of profile A (60 – 120 μmol/l). Mn concentrations tend to increase with depth in soil C and in the upper 60 cm of A, and were somewhat higher in 1981 than in later years, but distinct variations with time and depth are lacking.

Ammonium

Mean ammonium concentrations for whole soil profiles range from 60 (soil D) to 280 (soil B) μmol/l. Ammonium concentrations vary strongly with time and depth, with highest values just below the forest floor (0 cm depth). Except in summer 1983, concentrations are usually high (40 – 1000 μmol/l) in acidic surface horizons. In a few distinct periods, however, high ammonium concentrations appear throughout all or most of the four soil profiles: June 1981, September 1981, April 1982, August to December 1982, and, except in soil D, December to January 1983/84 (see Appendix 5).

Fluoride

Fluoride analysis by ion chromatography were never optimized, and the analytical data on this element (presented fully in the microfiche) should be regarded with caution. Therefore, F^- is not included in Table 34 or in Appendix 5. With confidence we can state, however, that F^- concentrations were generally below 30 μmol/l, except in soil A at 60 cm depth, were values were normally between 100 and 300 μmol/l. Perhaps these high values are due to contamination by the (single) ceramic cup involved here.

Chloride

As was discussed earlier, concentrations of Cl^- and Na^+ vary in a similar fashion, both showing seasonal variation with highs near the soil surface in the second part of the year, and both showing a gradual increase in concentration with depth. However, Cl^- peaks tend to appear later than Na^+ peaks, and Cl^- concentrations normally exceed those of Na^+ by a factor of about 1.5. Mean Cl^- concentrations for the whole soil profile are 0.45 (plots A and D), 0.34 (B) and 0.29 (C) mmol/l.

Nitrate

Nitrate is the dominant dissolved anion in all soils, with mean concentrations of 1.7 mmol/l (A), 1.4 mmol/l (B), 0.8 mmol/l (C) and 2.3 mmol/l (D). In plots A and D, nitrate concentrations in the soil solution are sometimes excessive, with values up to 8 mmol/l. The lowest nitrate concentrations in the soil profile are usually found at 0 cm, except in soil C, where the water percolating from the forest floor is relatively high in nitrate. The strong and well-defined seasonal variations in nitrate, and correlations with various cations were described above.

Sulfate

Mean ionic equivalent sulfate concentrations increase with depth from 0.5 or 0.7 at the soil surface to 1.5 mmol/l at 90 cm depth at plots A, B and D, and from 0.6 to 1.1 mmol/l at plot C. Sulfate concentrations vary seasonally with highest concentrations in summer and fall (surface soils) and in winter (subsoils). Except in soil A at 60 cm depth, sulfate concentrations are generally less well correlated than nitrate with concentrations of various cations, indicating that sulfuric acid is less important in mobilizing cations than nitric acid in our soils.

Phosphate

Phospate, presumably present as $H_2PO_4^-$, is usually observed in concentrations below 1 μmol/l. Concentrations decrease strongly with depth in the upper 10 to 20 cm of the soil profile. There seems to be a seasonal pattern with higher phosphorus concentrations throughout the profile in summer than in the rest of the year. As was mentioned before, phosphorus levels sometimes vary erratically, perhaps due to contamination.

Bicarbonate

Bicarbonate is present in the highest concentrations in the calcareous subsoils of plots A and D (2.4 mmol/l), and in somewhat lower concentrations at shallower depth in the soil at D (0.5 – 2 mmol/l).

Organic anions

Organic anion concentrations vary from about 0.1 to 0.3 mmol/l in the mineral soil. They are generally highest just below the forest floor (0.1 to 0.5 mmol/l), where organic anions make up 5 (A) to 20% (C and D) of the total anionic charge of the solution. At D, organic anion concentrations are higher than elsewhere (on account of the high pH and, consequently, high degree of dissociation), but there contribution to the total anionic charge is lower.

6.2.3 Concentrations of solutes in groundwater

Table 36 gives the means and standard deviations of the concentrations measured in monthly groundwater samples collected between March 16 1981 and May 1 1984 at each of the four monitoring plots.

Total electrolyte concentrations in the groundwater, as indicated by the electrical conductivity, are highest in plot B (932 μS/cm), lowest in $_{in}$C (464 μS/cm), and inter-

Table 36. Means (upper row) and standard deviation (lower row) of the electrical conductivity (EC, in $\mu S.cm^{-1}$) and solute concentrations ($\mu mol.l^{-1}$) in the groundwater of each of the plots A, B, C and D, 1981-1984. Values for ionic solutes are equivalent ionic concentrations (means of 39 samples).

	EC	pH	C_{org}	C_{in}	Si	K^+	Na^+	Ca^{2+}	Mg^{2+}	Fe^{2+}	Mn^{2+}	Al^{3+}	NH_4^+	F^-	Cl^-	NO_3^-	SO_4^{2-}	$H_2PO_4^-$
A	640	7.35	852	4050	138	21	407	5645	1060	6.4	3.7	2.3	18	7.8	413	1130	1900	0.63
	56	0.24	–	800	23	12	87	640	125	11.4	4.4	2.3	35	8.1	118	535	530	1.6
B	932	7.08	1130	3200	94	30	1285	7940	1320	10.5	15.1	6.9	22	22	1470	217	6400	0.52
	97	0.39	–	905	26	43	260	770	320	16.9	33.3	4.8	42	40	350	465	1810	1.1
C	464	4.44	495	715	505	36	820	1230	1815	510	64	28	44	119	915	100	3540	0.66
	37	0.43	–	420	455	44	103	161	269	375	120	6.7	63	60	195	205	314	2.62
D	785	7.27	1030	6880	104	14	405	7770	1220	11.1	12.7	26	23	16	395	31	3070	0.32
	53	0.24	–	1330	20	7.5	52	620	86	16.6	22.4	13	35	55	97	48	740	0.64

mediate in A and D. The standard deviations are in the order of 10% or less of the mean, showing that the total electrolyte contents do not vary strongly with time. The same holds for the bulk ionic components C_{in} (mainly bicarbonate, at least at A, B and D), Na^+, Ca^{2+}, Mg^{2+}, Cl^- and SO_4^{2-}, where the standard deviation is generally within 25% of the mean. At plots A and D, the composition of the groundwater (sampled at 150 cm depth) is rather similar to that of the soil solution at 90 cm depth, except for the nitrate concentration, which is distinctly lower in the groundwater. At plots B and C, the concentrations of most solutes, except potassium, aluminium and nitrate, are much higher in the ground water than in the soil solution at 90 cm depth. The higher pH, the higher concentrations of Ca^{2+}, Mg^{2+}, C_{in}, and the lower concentrations of Al^{3+} and K^+ in the ground water at B can be explained by precipitation reactions at high pH due to the presence of calcium carbonate in the substratum. The same trends can be observed in the soil solutions at plot A, where $CaCO_3$ appears at much shallower depth. At C no $CaCO_3$ was observed within 2.5 m below the surface, and the groundwater at 150 cm is indeed disctinctly acidic. The relatively high Ca^{2+} and Mg^{2+} levels in the groundwater at C may be due to dissolution of easily weatherable silicate minerals in the substratum, but may partly reflect mixing with groundwater from calcareous zones. Why the concentrations of Na^+, Cl^- and SO_4^{2-} at B and C are about 2 – 3 times higher in the groundwater than in the surface water is difficult to explain. As will be discussed below, the high sulfate concentrations may be due to oxidation of pyrite, which is present in small concentrations in the zone below the deepest groundwater table. If most of the Na^+ and Cl^- is of atmospheric origin, which is very probable, there must be some concentration process operating below 90 cm depth in the soil. Because few if any roots are available for evaporative concentration of the deeper water, this is difficult to understand.

The low nitrate concentrations in the groundwater can be explained by denitrification of the nitrate leached down from the soil. Dissolved organic carbon, which is present in fairly high concentrations in the groundwater could serve as a reducing agent. However, correlations between NO_3 and organic C in the groundwater samples from each plot were not significant ($-0.2 < r < 0.3$), indicating little effect of total dissolved organic C on nitrate. At plot A nitrate concentrations in the ground water showed a strong seasonal variation, with an increase in nitrate starting in the beginning of the year, followed by a decrease in the summer (Fig 32). Sulfate showed the reverse pattern, as evinced by the strong negative correlation ($r = -0.67$) between the SO_4^{2-} and NO_3^- concentrations. The increase in nitrate can be explained by the arrival, at 150 cm depth, of nitrate leached from the soil with the drainage water. The subsequent decrease in nitrate and increase in sulfate by denitrification by pyrite, e.g. according to:

$$5FeS_2 + 15NO_3^- + 5HCO_3^- \rightarrow 5FeOOH + 7.5N_2 + 10SO_4^{2-} + 5CO_2$$

The changes in (equivalent ionic) concentrations of nitrate and sulfate over each of the three annual cycles at plot A are in the same order of magnitude (in mmol/l:

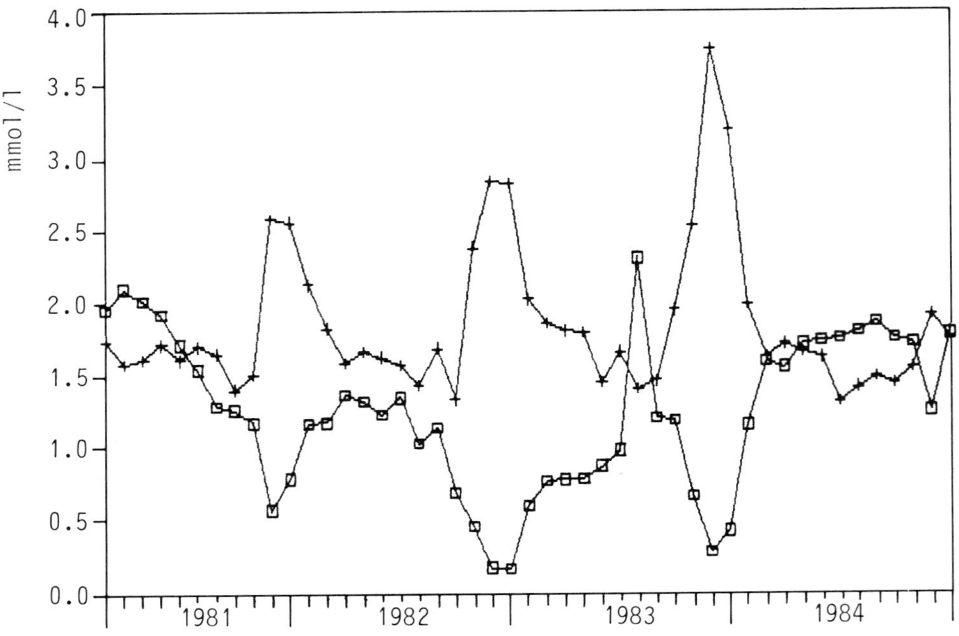

Figure 32. Contents of nitrate (squares) and sulfate (crosses) in the ground water (ionic equivalent concentrations in mmol/l) of plot A as a function of time between March 1981 and December 1984.

resp 1.5 vs. 1.0 in 1981, 1.2 vs. 1.2 in 1982, and 2.0 vs. 2.3 in 1983). This is in reasonable agreement with the stoichiometry of this reaction. At the other plots, nitrate concentrations were much lower, often with undetectable concentrations (< 1 or 10 mmol/m^3) around January-February, and somewhat erratic peak values during summer. Sulfate concentrations in the groundwater at B, C and D were much higher than at A, without a clear seasonal variation. Possibly, the lower nitrate concentrations and higher sulfate concentrations in the ground water at plots B, C and D are caused by a more rapid and complete progress of the same reaction, e.g. because more pyrite is available for the process.

The large standard deviations of the elements present in trace concentrations (Mn^{2+}, Al^{3+}, NH_4^+, $H_2PO_4^-$ and, exept plot C, Fe^{2+}), may be in part due to analytical problems and contamination. The relatively high concentration of dissolved Fe^{2+} at C must be due to chemically reduced conditions at a relatively low pH. Considering the pH of the groundwater at A, B and D the Al^{3+} concentrations appear to be high. Probably most of this aluminum is present in polynuclear form.

6.3 Input-output budgets

6.3.1 Water fluxes

The SWATRE model was calibrated primarily by fitting calculated hydraulic potentials to the measured values. Appendix 6 shows various measured and calculated waterfluxes, in cm, over monthly periods for each of the three hydrological years considered: infiltration into the soil (= throughfall), transpiration from the soil, and the downward (−) or upward waterfluxes at 10, 20, 40, 60 and 90 cm depth in the soil. Throughfall fluxes refer to means of the seven collectors per plot. Table 37 shows the main water balance terms for all plots for the calibration and validation year and the hydrological year 1983-1984. At the start of each hydrological year the soil profile is in equilibrium with the groundwater, and any net storage differences are due to differences in groundwater level. In 1981 the net evapotranspiration reduction was negligible for all plots except C, which has the lowest water holding capacity. In 1982 the reduction in evapotranspiration was considerable except for plot D. Plot D rarely shows a water deficit, possibly due to a more direct uptake of infiltrated water as a result of a concentration of rootmass in the topsoil.

Although in woodlands evapotranspiration tends to be relatively high due to evaporation of water intercepted by the canopy, in our case it was depressed by reductions of transpiration due to recurrent infestation of the caterpillar *Tortrix viridiana*

Table 37. Main annual water balance terms (mm) for plot A to D from April 1, 1981 to March 18, 1984. First and last dates of the hydrological years are:
I = 4.1.1981-3.22.1982, II = 3.22.1982-3.22.1983, and III = 3.22.1983-3.18.1984 (month.day.year).

plot	period	gross precip.	pot. evapot.	actual evapot.	actual interc.	actual soil evap.	actual transp.	actual drain-age	actual storage change
A	I	643	480	471	128	48	294	182	-10
	II	659	541	505	140	48	317	148	6
	III	817	545	563	197	67	299	242	12
B	I	643	480	473	126	53	294	166	4
	II	659	541	493	146	50	297	152	14
	III	808	545	524	184	66	274	277	7
C	I	643	480	453	125	53	275	249	-59
	II	659	541	470	140	53	277	195	-6
	III	812	545	510	185	67	258	297	5
D	I	643	480	473	121	53	299	202	-32
	II	659	541	520	147	52	321	144	-5
	III	829	545	560	157	67	336	264	5

Table 38. Reduction and enhancement terms for the transition of potential to actual evapotranspiration at plots A to D in each of the three hydrological years (mm).

plot	period	total	inter-ception	Tortrix viridiana	water deficit	surface conductivity
A	I	-9	90	-33	-12	-58
	II	-36	102	-37	-37	-64
	III	18	165	-40	-67	-37
B	I	7	87	-33	-8	-53
	II	-4	105	-37	-48	-63
	III	-21	147	-40	-90	-38
C	I	-27	84	-33	-25	-53
	II	-71	98	-37	-72	-60
	III	-35	149	-40	-108	-36
D	I	-8	79	-33	-1	-53
	II	-21	100	-37	-23	-61
	III	15	124	-40	-33	-36

(oak leaf roller) in May-June, and reduction of soil evaporation due to low conductivity of the surface soil. Together with the soil water deficit, these factors almost balanced the evaporation of intercepted water (Table 38). So annual evapotranspiration at our small, relatively isolated woodland (with appreciable advection) should be similar to that from large, extended forest areas (without advection) with a negligible soil water deficit. Large forests under similar climatic conditions evaporate 450 – 550 mm (Rutter, 1968), which is in the same range as the values calculated for our sites. In the big ($20 \times 20 \times 3$ m^3) lysimeters at Castricum, in the western part of the Netherlands, the average annual evapotranspiration for a mixed oak-birch forest between 1955 and 1964 was 500 mm (Mulder, 1983). Similarly, the GELGAM model (De Laat, 1980) gave an average evapotranspiration of 472 mm between 1971 and 1980 for deciduous woodland in our region (Van Bakel and Van de Nes, 1984).

Simulated soil water fluxes are illustrated graphically in Fig. 33. Water net transport below 40 cm depth is practically zero between June and September. Drainage increases gradually from autumn to spring, with a delay for one or two months at 60 and 90 cm depth. Flux patterns for plot A, B and C are very similar. The systematically higher transpiration at D is reflected by the small water fluxes in summer in that plot. The water flux patterns for 1981-1982 and 1982-1983 are rather similar, except for the occurrence of the rainstorm in August 1982, which apparently infiltrated to a depth of 90 cm. The hydrological year 1983-1984 had an extremely wet spring, followed by a dry summer.

An important question is the effect of a possible underestimation of the precipitation (see paragraph 6.1) on the calculated water fluxes. If precipitation was underestimated indeed, interception (= precipitation minus throughfall) would be unde-

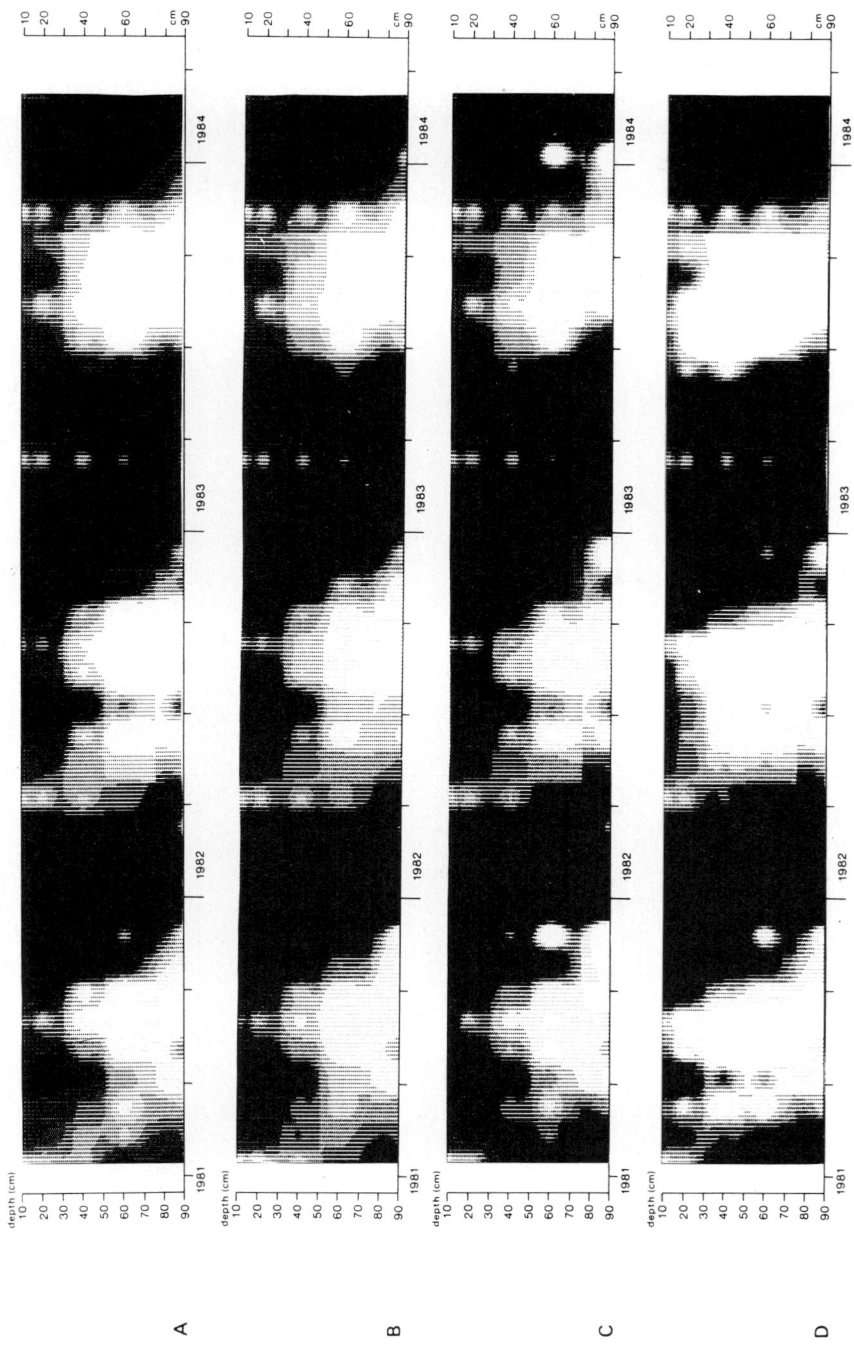

Figure 33. Simulated monthly soil water fluxes as a function of time and depth from 4 – 1 – 1981 to 3 – 18 – 1984 at plots A, B, C and D. Darker shading indicates increasing downward fluxes from + 50 (white) to – 100 mm (black).

restimated too, leading to an overestimation of the transpiration, and a decreased drainage flux. In practice, however, the variation in calculated transpiration with hydraulic potential under quite dry conditions, is much smaller than the corresponding variation in potential tranpiration (Postma, 1986). Moreover, the underestimation of the transpiration will only be important in summer, when there is generally a water deficit, so that higher rainfall rates would have little if any effect on water movement in the lower part of the rootzone and in the subsoil. The winter period is far more important for the transport of water and chemicals. In winter, transpiration is very small, and percolation rates are determined largely by infiltration (i.e. measured throughfall) which is not subject tot a potential systematic error. So in conclusion, an underestimation of the precipiation, if any, should not result in an appreciable underestimation of the drainage rates.

6.3.2 Solute fluxes

Table 39 summarizes the annual solute fluxes in precipitation and throughfall for each of the hydrological years, and in stemflow for the three-year period. The relative orders of magnitude of the fluxes of different solutes within a type of water (i.e. precipitation, throughfall or stemflow are similar to those of concentrations, alrea-

Table 39. Annual solute fluxes (equivalent kmol/ha.yr) and water fluxes (mm) in precipitation, throughfall and stemflow.

	H^+	C_{org}	K^+	Na^+	Ca^{2+}	Mg^{2+}	Fe^{2+}	Mn^{2+}	NH_4^+	Cl^-	NO_3^-	SO_4^{2-}	$H_2PO_4^-$	water
Precipitation														
1981/82	0.37	1.40	0.03	0.80	0.30	0.18	0.020	0.006	1.23	0.79	0.48	1.35	0.012	643
1982/83	0.13	1.64	0.04	0.87	0.40	0.21	0.020	0.005	1.01	0.94	0.48	1.13	0.008	659
1983/84	0.08	1.29	0.08	0.95	0.47	0.24	0.013	0.007	1.33	0.96	0.66	1.43	0.015	817
Throughfall A														
1981/82	0.11	8.48	0.87	1.00	0.61	0.41	0.016	0.056	2.30	1.27	0.84	3.42	0.05	514
1982/83	0.12	6.27	0.94	1.21	0.75	0.43	0.020	0.063	2.61	1.67	0.91	2.84	0.05	519
1983/84	0.03	9.61	1.45	1.18	0.76	0.60	0.023	0.079	2.93	1.35	1.09	3.24	0.11	620
Throughfall B														
1981/82	0.17	8.44	0.85	0.95	0.57	0.38	0.016	0.025	3.18	1.17	0.95	3.18	0.06	517
1982/83	0.14	7.17	1.03	1.07	0.62	0.39	0.019	0.033	2.73	1.55	0.87	2.60	0.08	513
1983/84	0.03	10.2	1.58	1.20	0.74	0.62	0.015	0.043	3.06	1.48	1.24	3.19	0.13	624
Throughfall C														
1981/82	0.23	6.22	0.59	0.85	0.50	0.36	0.018	0.036	2.47	1.19	0.83	2.55	0.06	518
1982/83	0.13	5.89	0.78	0.96	0.63	0.41	0.014	0.043	2.20	1.36	0.75	2.20	0.08	519
1983/84	0.04	8.18	1.28	1.06	0.66	0.57	0.014	0.055	2.34	1.25	0.96	2.46	0.12	627
Throughfall D														
1981/82	0.09	10.7	1.02	1.25	1.26	0.64	0.016	0.045	3.72	1.84	0.98	4.34	0.04	522
1982/83	0.07	9.26	1.11	1.62	1.35	0.66	0.020	0.047	3.02	2.53	0.92	3.53	0.07	512
1983/84	0.02	12.0	1.45	1.59	1.40	0.82	0.017	0.058	3.65	2.13	1.17	4.15	0.12	672
Stemflow (means for 1981/83)														
A	0.00	0.29	0.03	0.05	0.02	0.02	0.00	0.002	0.21	0.06	0.03	0.22	0.00	10.4
B	0.00	0.35	0.03	0.04	0.02	0.01	0.00	0.001	0.23	0.06	0.02	0.23	0.00	10.4
C	0.00	0.25	0.03	0.04	0.02	0.01	0.00	0.001	0.19	0.05	0.02	0.19	0.00	9.6
D	0.00	0.28	0.02	0.05	0.03	0.02	0.00	0.001	0.19	0.06	0.02	0.20	0.00	10.5

dy described in paragraph 6.2. Solute fluxes were largest in the 1983/84 hydrological year (which was also the wettest year), but the annual values were generally similar within 20 – 30%. The relative importance of dry deposition and leaching from, or uptake by, above-ground parts of the vegetation will be discussed below, in order to obtain an estimate of the total atmospheric input of various elements in the ecosystems considered.

A distinct trend with time over the three year period may be present for the flux of H^+, which decreases very strongly in all types of deposition. The decrease may be related to contamination by spray due to sprinkling irrigation (with groundwater high in calcium bicarbonate) of the neighbouring pasture land to the west and southwest of the monitoring plots, a practice that started in 1983.

As with concentrations, fluxes in throughfall were generally highest at D and lowest at C, with intermediate values at A and B. Based on triplicate throughfall values obtained in 1981, only plots A and C showed significant differences ($P \leq 0.05$) in throughfall inputs for most elements (exceptions were Fe^{2+} and NO_3^-). Only for organic C, Ca^{2+} and Cl^- throughfall fluxes at D were significantly higher than those at the other plots.

Stemflow inputs on an area basis were small and generally less than 5 to 10% of the throughfall inputs. Of course, stemflow may cause high, strongly localized, inputs at the base of trees, but these will not be considered further in this study.

Fluxes of most elements strongly increase during passage of rain water through the tree canopies. The ratio of solute fluxes in precipitation to those in throughfall, or deposition ratios, DR (Parker, 1983), provides a measure of this increase. Deposition ratios are shown in Table 40. Deposition ratios are always smaller than concentration ratios (see paragraph 6.2.1) because part of the precipitation water is intercepted by the canopy. In the canopy, free H^+ is partially neutralized (DR < 1). Strongest neutralization of H^+ is observed in summer at plot D. For all other elements except iron, the net effect of dry deposition, assimilation of solutes in the ca-

Table 40. Mean deposition ratios (DR = throughfall flux/precipitation flux) for plots A plus B plus C, and for plot D in the vegatation period (15 April to 1 December, 'summer') and in the leafless period ('winter'). For comparison, mean annual DR values for many different forest ecosystems are shown. Values for the 'Oude Maat' are 4-year means (January 1981 to December 1984).

	H^+	C_{org}	K^+	Na^+	Ca^{2+}	Mg^{2+}	Fe^{2+}	Mn^{2+}	NH_4^+	Cl^-	NO_3^-	SO_4^{2-}	$H_2PO_4^-$
'Oude Maat' A, B and C													
summer	0.46	5.7	25	0.94	1.5	2.4	0.97	6.3	1.9	1.4	1.8	1.8	7.1
winter	0.68	2.2	6.2	1.7	1.4	1.6	0.49	4.2	2.1	1.4	1.5	2.4	3.7
'Oude Maat' D													
summer	0.25	8.5	30	1.4	3.8	4.1	1.1	6.1	2.6	2.5	2.3	2.6	7.6
winter	0.93	4.0	9.2	2.4	3.3	3.1	0.66	7.1	3.7	2.7	2.3	4.1	4.2
Average from literature (Parker, 1983)													
	–	–	11	2.4	2.3	4.0	–	–	1.6	3.0	1.3	2.3	3.9

nopy, and canopy leaching, is a net increase in the solute flux reaching the ground surface (DR > 1). Deposition ratios for other elements have the same order of magnitude as those observed elsewhere (Table 40), but values for K^+, N^+ and $H_2PO_4^-$, are relatively high and those for Na^+, Mg^{2+} and Cl^- are on the low side. Some elements have distinctly higher deposition ratios in the vegetation period than when the trees are leafless (viz. organic carbon, K^+, Mg^{2+}, Fe^{2+} and $H_2PO_4^-$). For organic carbon, K^+, Mg^{2+} and $H_2PO_4^-$, the increase in flux taking place while water passes the canopy can be attributed almost entirely to canopy leaching, i.e. extraction of solutes from interior parts of the vegetation by rain water (Parker, 1983). Mainly because distrinctly higher Ca^{2+} deposition ratios are observed at the calcareous site D (where trees clearly take up more Ca^{2+}) than at the plots with acidic soils, the increase in the flux of calcium during passage of precipitation water through the canopy must be attributed mainly to leaching as well, although local atmospheric sources (dust from chicken barns and spray from irrigation water) may play a role too. High deposition ratios of Mn^{2+} in both winter and summer indicates that this element may appear in throughfall water by leaching of both bark and leaves.

For Na^+, Cl^-, NH_4^+ and SO_4^{2-}, concentrations in throughfall as well as deposition ratios are higher in winter than in summer, in spite of stronger evaporation of intercepted water during summer. Because for the same elements gaseous (SO_2 and NH_3) and aerosol (NaCl) concentrations tend to be higher in winter too (Van Aalst & Van Diederen, 1983; Annual reports of the joint KNMI/RIVM project 'Chemical composition of precipitation over the Netherlands'), most of the increase in the flux of these substances during passage of water along above-ground parts of the vegatation must be ascribed to washing off the dry-deposited material. The similar magnitude and high correlation of the concentrations of Na^+ and Cl^- in throughfall and stemflow ($r = 0.87$) also suggest an important deposition of NaCl in aerosol form. The fact that concentrations of NH_4^+, SO_4^{2-}, Na^+ and Cl^- in throughfall are significantly higher in the leafless period than in the vegetation period (Table 33) suggests that the presence of a leafy canopy (which would enhance dry deposition) is less important than other factors, including gaseous and aerosol concentrations (which are higher in winter than in summer) in determining rates of dry deposition.

It is unlikely that significant quantities of nitrate are leached from above-ground parts of trees (Parker, 1983), so most of the difference in nitrate flux between precipitation and throughfall must probably be ascribed to washing off dry-deposited NO_3^-. Distinctly higher NO_3^- concentrations in summer than in winter may indicate that aerosol nitrate and gaseous HNO_3, which have higher concentrations in the atmosphere in summer than in winter (Van Aalst et al., 1983) are more important as a source for dry-deposited nitrate than NO_x (which shows highest atmospheric concentrations in winter). Favoured by higher temperatures, nitrification of ammonium, both on leaf surfaces and in the throughfall collectors, may also have contributed tot the relatively high nitrate concentrations in summer.

Our throughfall fluxes of NH_4^+ (about 3 kmol/ha.yr) and of NO_3^- (about 1 kmol/ha.yr) are far higher than any reported in two recent literature reviews (Parker, 1983; VDI, 1983), and about an order of magnitude higher than the mean values of all literature values (0.3 $\pm$0.5 kmol/ha.yr for NH_4^+ and 0.1 $\pm$ 0.2 kmol/ha.yr for NO_3^-, Parker, 1983). The high deposition of ammonium is related to high atmospheric concentrations of ammonia from intensive animal husbandry in the general area. The good correlation between the concentrations of ammonium and sulfate in throughfall and stemflow water, and the nearly equivalent concentrations of those ions (Fig. 30) must be ascribed to mutually enhanced dry deposition of NH_3 and SO_2 (Van Breemen et al., 1982; Buysman et al., 1984; Adema et al., submitted for publication). Our throughfall fluxes of N and S are of the same order as those estimated in the Netherlands for dry + wet deposition by other procedures (Onderdelinden et al., 1984; Van Aalst et al., 1983). The high ammonium sulfate concentrations on leaf surfaces may have contributed to the relatively high throughfall flux of K^+ at our sites, by a leaching effect of ammonium sulfate on K^+ in leaves (Roelofs et al., 1985). This points to another problem in trying to partition throughfall fluxes in contributions from dry and from wet deposition: the possibility of foliar uptake of atmospherically derived substances. If ionic substances are involved, uptake involves ion exchange, and at least the order of magnitude can be guessed. For instance, in the case of the $NH_4^+ - K^+$ exchange referred to above, the amount of NH_4^+ taken up can never exceed the amount of K^+ found in the throughfall. Na^+ too may have some leaching effect on cations in leaves, as suggested by the fact that deposition ratios for Na^+ are similar for those of Cl^- in winter, but lower in summer. The amount of atmospherically derived H^+ exchanged for base cations in leaves can be estimated as follows. On an equivalent ionic basis, the increase in the flux of nitrate plus sulfate in meteoric water passing the tree canopy invariably exceeded the corresponding increase in the flux of ammonium. This difference can be accounted for by H^+ from sulfuric and nitric acid that was not neutralized by ammonia, and can be considered as dry-deposited H^+. Such excess sulfuric plus nitric acid may be the result of:

1. higher deposition of SO_2 and NO_x than of NH_3,
2. partial nitrification of ammonium in the canopy,
3. uptake of ammonium in the canopy, or any combination of these cases.

The total H^+ deposited on the canopy, dry deposited H^+ plus H^+ deposited in the open collectors, is designated as H_t. Table 41 gives three-year (April 1981 – April 1984) average annual fluxes of major solutes and water, including H_t. H_t was calculated as follows:

$$H_t = SO_4^{2-}(thr) - SO_4^{2-}(p) + NO_3^-(thr) - NO_3^-(p) - (NH_4^+(thr) - NH_4^+(p)) + H^-(p),$$

where chemical symbols refer to equivalent ionic fluxes, and (thr) and (p) stand for throughfall and precipitation, respectively. Actual H^+ fluxes in canopy throughfall were much lower than H_t. Most of the H^+ deposited was removed in the canopy by exchange with other cations and thus accounted for part of the cation leaching discussed above. These acid inputs will eventually influence the soil system when the leached cations are replenished by root uptake (Ulrich, 1983).

Table 41. Three-year (April 1981-April 1984) average annual fluxes of major solutes (kmol/ha.yr) and water into collectors in the open field (P) and under tree canopies in each of the monitoring plots A, B, C and D. Underscored values are considered to represent inputs from the atmosphere. Phosphorus is assumed to be present as $H_2PO_4^-$. H_t refers to the total (wet + dry deposited) free H^+ input to the canopy.

	H^+	K^+	Na^+	Ca^{2+}	Mg^{2+}	Mn^{2+}	NH_4^+	Cl^-	NO_3^-	SO_4^{2-}	$H_2PO_4^-$	H_t	water (mm)
P	0.19	$\underline{0.05}$	0.88	$\underline{0.39}$	$\underline{0.21}$	$\underline{0.006}$	1.19	0.90	0.54	1.30	$\underline{0.01}$	–	706
A	0.09	1.08	$\underline{1.13}$	0.70	0.48	0.066	$\underline{2.95}$	$\underline{1.43}$	$\underline{0.95}$	$\underline{3.17}$	$\underline{0.07}$	$\underline{0.71}$	551
B	0.11	1.15	$\underline{1.07}$	0.64	0.46	0.034	$\underline{2.99}$	$\underline{1.40}$	$\underline{1.02}$	$\underline{2.99}$	$\underline{0.09}$	$\underline{0.56}$	551
C	0.13	0.89	$\underline{0.96}$	0.60	0.45	0.045	$\underline{2.34}$	$\underline{1.25}$	$\underline{0.85}$	$\underline{2.40}$	$\underline{0.09}$	$\underline{0.45}$	554
D	0.06	1.19	$\underline{1.49}$	1.34	0.71	0.050	$\underline{3.46}$	$\underline{2.17}$	$\underline{1.02}$	$\underline{4.00}$	$\underline{0.08}$	$\underline{1.10}$	569

With the methods applied here, there is no way to quantify the amounts of uncharged gaseous substances (NH_3, NO_x, HNO_3 and SO_2) assimilated by leaves, which may be particularly important for nitrogen. Therefore, actual dry deposition of N and S may still be higher than the difference between throughfall and precipitation fluxes.

Based on the considerations above, we have assumed that the differences between precipitation and throughfall fluxes for K^+, Ca^{2+}, Mg^{2+}, Mn^{2+} and $H_2PO_4^-$ are due entirely to canopy leaching, and total atmospheric inputs of these elements to the forest are equal to those observed in the precipitation collectors in the open field. Because tree canopies are presumably more efficient in scavenging aerosols and dust than the bulk collectors, dry deposition of those elements is probably underestimated (and leaching overestimated). On the other hand, for N, S, Cl^- and Na^+, differences between throughfall and precipitation fluxes have been attributed to dry deposition only. For Na^+ and Cl^- actual dry deposition may have been less, because cycling through the vegetation could have contributed to the increased throughfall flux. For N total atmospheric deposition may have been higher than the throughfall flux as a result of unseen foliar uptake of gaseous N-compounds.

Underlined data in Table 41 indicate total atmospheric inputs as estimated above. Total inputs of most elements to the plots were 2 – 10% higher due to the contribution of stemflow. However, stemflow input to the soil is confined to the immediate vicinity of the trees, and has little influence on the bulk of the soil system, for which the below-ground solute fluxes have been determined. In an earlier publication (Van Breemen et al., 1986) the same table was presented with precipitation fluxes, rather than throughfall fluxes, for Na^+ and Cl^- as the best estimate for total atmospheric deposition. In that publication, the increased flux of Na^+ and Cl^- in throughfall was attributed wholly to cycling because:

(1) soil solution fluxes of Na^+ and Cl^- tended to decrease with depth in the root zone (see Table 47), and

(2) soil solution fluxes of Cl^- below the rootzone were similar to the precipitation inputs.

Here, the emphasis is placed on dry deposition of NaCl, based on the data on seasonal variation of NaCl in throughfall. The truth must lie somewhere between those two sets of values. The available data do not permit to partition the increased throughfall flux of NaCl over cycling and dry deposition.

The difference in inputs of S and N among the plots, with lowest values at C, highest values at D, and intermediate values at A and B, are very consistent from season to season and from year to year (Table 39). At least for A and C the differences in solute inputs are statistically significant, and may result from differences in exposure to atmospheric pollutants as a result of differences in exposition to prevalent winds, distances to pollution sources, and differences in canopy stucture and density. Partly, however, the differences may be due to biased sampling of throughfall. For instance, the high values of S and N inputs at D are caused mainly by collector D3 (see paragraph 6.2.1).

6.3.3 Leaf litter inputs and nutrient cycling by trees

Table 42 gives means and standard deviations of the tree litter supplied to the soil surface from April to April in 1980-81 and 1981-82. Oak dominates litterfall at all plots, but alder and poplar are almost equally important at plot D, while birch supplies significant amounts of litter at plot A. 'Total litter' includes unrecognizable material, and therefor exceeds the summed values of the litter fluxes attributable to various species ('total species'). Total litter is highest at plot B, and lowest at plot A.

Tables 43 and 44 give the concentrations of elements in various types of litter. At plot D, with four species of trees, differences in chemical composition of the litter of different species can be evaluated. The contents of Na, Cl, P and S are rather similar

Table 42. Annual litter fall at each of the four plots. Upper rows give means for six collectors, the second row gives standard deviations (ton/ha).

plot	oak	birch	poplar	alder	total species	1980/81 total litter	1981/82 total litter
A	2.54	1.83	0.0	0.0	4.37	5.32	5.29
	0.59	0.28				0.63	0.41
B	5.07	0.46	0.0	0.0	5.51	6.69	6.31
	0.95	0.58				0.73	0.95
C	4.91	0.07	0.0	0.0	4.98	5.67	6.13
	0.74	0.13				0.80	1.58
D	1.48	0.54	1.26	1.47	4.75	5.70	5.92
	0.68	0.28	0.89	0.46		0.77	0.73

Table 43. Concentrations of elements in tree litter from oak, birch, poplar and alder growing in calcareous soil at plot D, collected from May 1980 to May 1981. Means of six annual composite samples (mmol/kg(±)). If the concentration for a given element in a certain type of sample varies significantly (P < 0.05) among the four tree species, this is indicated by different letters printed behind the mean value.

	K	Na	Ca	Mg	Fe	Mn	Al	Cl	S	P	N	SO_4
Oak	88a	40	296a	61ab	5a	14 b	5a	31	63	29	1459 b	7
Birch	78a	26	337ab	63a	7a	15 b	10 b	47	–	25	1353 b	4
Poplar	285 c	35	535 c	84 b	5a	5a	5a	62	79	36	1181a	5
Alder	131 b	21	450 b	84 b	9 b	5a	10 b	41	74	25	1823 c	10

Table 44. Concentrations of elements in various types of tree litter (oak, birch, total) in each of the four plots A, B, C and D, collected from May 1980 to May 1981. Means of six annual composite samples (mmol/kg(±)). If the concentration for a given element in a certain type of sample varies significantly (P < 0.05) among the three or four plots, this is indicated by different letters printed behind the mean value.

	K	Na	Ca	Mg	Fe	Mn	Al	Cl	S	P	N	SO_4
Oak												
A	133	17	166 b	58	7 c	21 c	8	25	67 b	35	1529	17 b
B	110	27	148a	53	5abc	8a	6	29	48a	32	1409	14ab
C	85	13	151ab	62	3a	15 b	5	21	67 b	34	1349	8a
D	88	40	296 c	61	5 b	14 b	5	31	63 b	29	1459	7a
Birch												
A	126 b	15	171a	62	6	26 b	7a	41	69	26	1374	15
B	162 b	33	159a	63	9	11a	12 b	49	26	25	1270	5
D	78a	26	337 b	63	7	15a	10ab	47	–	25	1353	4
Total tree litter												
A	118	28a	163a	53a	10	19 c	12	23a	64	34	1361	14
B	144	21a	174a	56a	8	9a	11	23a	59	42	1510	15
C	126	28a	168a	62ab	7	15 b	9	21a	58	43	1467	14
D	196	55 b	346 b	68 b	9	9a	10	44 b	73	34	1521	23

for the various litter types. Poplar litter has significantly more K and Ca, and less N than litter of other species. Alder had by far the highest N contents, and is intermediate between poplar and oak or birch in terms of K and Ca. Oak and birch have litter of very similar chemical composition. Comparing litter of a certain species from different plots may indicate the effect of soil on litter composition and nutrient cycling. Table 44 shows that both oak and birch growing on calcareous soil (plot D) have litter containing significantly more Ca than trees growing on the plots with non-calcareous soils. The relatively high Mn concentrations in the soil solution at A may be the cause of the significantly higher Mn contents in tree litter at this plot. Note that Al contents in tree litter do not reflect in any way the large differences in water-

Table 45. Relative fluxes of elements in total tree litter (expressed in % of the mean flux observed in all 24 litter collectors) per plot for 1980-1982. Means that vary significantly ($P < 0.05$) among the four plots, are indicated by different letters printed.

	K	Na	Ca	Mg	Fe	Mn	Al	Cl	S	P	N	SO_4
A	79a	84a	70a	83a	103ab	139 c	110ab	85a	86a	89a	89a	87
B	110 b	98ab	89 b	103 b	106 b	74a	114 b	92ab	106ab	120 b	112 b	102
C	91ab	80a	73ab	104ab	81a	106 b	69a	87a	98ab	105ab	96ab	101
d	121 b	139 b	168 c	110 b	109 b	81ab	107 b	136 b	110 b	89a	104ab	111

Table. 46. Element cycling through the tree vegetation at each of the monitoring plots. Values are fluxes in kmol(±)/ha.yr, assuming valencies of 1 for N (NO_3^- or NH_4^+) and P ($H_2PO_4^-$), 2 for S and 3 for Al. Total uptake of N includes storage in the forest floor.
--: not known, but assumed tot be negligible.

		H	K	Na	Ca	Mg	Mn	Al	Cl	S	P	N
A.	litter	--	0.65	0.11	1.75	0.57	0.23	0.24	0.14	0.56	0.18	7.54
	leaf leach.	-0.62	1.03	--	0.31	0.27	0.06	--	--	--	0.06	--
	net uptake	--	0.07	0.00	0.09	0.03	0.00	0.00	--	0.00	0.01	0.36
	tot.uptake	--	1.75	0.11	2.15	0.87	0.29	0.24	0.14	0.56	0.25	7.20
B.	litter	--	0.91	0.12	2.23	0.71	0.12	0.25	0.14	0.70	0.26	9.50
	leaf leach.	-0.49	1.10	--	0.25	0.25	0.03	--	--	--	0.08	--
	net uptake	--	0.25	0.02	0.37	0.12	0.00	0.00	--	0.00	0.03	1.19
	tot.uptake	--	2.26	0.14	2.85	1.08	0.15	0.25	0.14	0.70	0.37	9.99
C.	litter	--	0.75	0.11	1.84	0.71	0.18	0.15	0.14	0.64	0.22	8.10
	leaf leach.	-0.32	0.84	--	0.21	0.24	0.04	--	--	--	0.08	--
	net uptake	--	0.29	0.02	0.41	0.13	0.00	0.00	--	0.00	0.03	1.34
	tot.uptake	--	1.88	0.13	2.46	1.10	0.22	0.15	0.14	0.64	0.33	8.74
D.	litter	--	1.01	0.21	4.22	0.76	0.14	0.23	0.22	0.72	0.19	8.77
	leaf leach.	-1.06	1.14	--	0.95	0.50	0.04	--	--	--	0.07	--
	net uptake	--	0.23	0.00	0.99	0.08	0.00	0.00	--	--	0.04	1.05
	tot.uptake	--	2.38	0.21	6.16	1.34	0.18	0.23	0.22	0.72	0.30	9.82

soluble Al among the various plots, indicating that the mobility of Al within these trees is low, and that any toxic effects of Al would be confined to parts other than the leaves.

Table 45 gives the element fluxes per plot relative to the mean for the four plots, which allows a comparison of the four plots in terms of element cycling. Mainly owing to lower litter production, the stand at plot A cycles significantly less K, Ca, Mg, P and N than that at plot B. At plot D, the flux of Ca in litter is about twice as high as at the other plots.

Table 46 shows, for each of the four plots, the annual fluxes of elements in leaf litter (means for 1980/81 + 1981/82), the fluxes of elements leached from the tree canopy (calculated as the difference between throughfall input and total atmospheric input, Table 41), and the annual net uptake in wood, bark and roots. The annual net uptake was obtained from the measurements by De Visser (1986). Total uptake of N was calculated assuming that 0.7 kmol/ha.yr of N accumulates in the forest floor. The value of 0.7 is derived from comparisons of recent (1985) measurements of the N-pools in the forest floor by Winkels (1985) with data for 1960 by Minderman (1981). Only a small fraction of the annual uptake from the soil is stored in woody parts and bark ('net uptake'): most is returned to the soil as leaf litter (especially Ca, Mg, Mn, S, P and N) and canopy leachate (especially K).

6.3.4 Solute fluxes in the soil

Appendix 7 gives annual solute fluxes at the depths of the soil solution collectors, i.e. at 10, 20, 40, 60 and 90 cm depth, as calculated by summing monthly fluxes obtained from waterfluxes by SWATRE (paragraph 6.3.1) multiplied by measured soil solution concentrations. Fluxes of organic anions plus bicarbonate were estimated from the cation/anion charge balance (A^-). Solute fluxes in the individual hydrological years were generally within 30% of the tree-year average annual values shown, for selected soil depths, in Table 47. Lack of sufficient replicated samples in the monitoring programma prevented a good estimate of the uncertainty associated with

Table 47. Three-year mean drainage fluxes of water (mm/yr) and solutes (equivalent ionic fluxes in kmol(±)/ha.yr) at 0, 10, 60 and 90 cm depth. Values at 0 cm depth were either estimated from 1983-84 chemical analyses of the 0-cm soil solutions (a), or from data on litterfall and throughfall inputs, assuming H^+ and organic anion data according 1983-84 0-cm soil solution analysis (b).

plot:	depth:	H^+	K^+	Na^+	Ca^{2+}	Mg^{2+}	Al^{3+}	NH_4^+	Cl^-	NO_3^-	SO_4^{2-}	$H_2PO_4^-$	A^-	water
A.	0 cm(a)	1.38	2.04	1.35	2.64	1.29	0.42	2.12	1.68	5.49	3.09	0.17	0.60 #	
	0 cm(b)	1.38	1.73	1.24	2.45	1.05	0.24	4.40	1.57	6.34	3.73	0.25	0.60 #	551
	10 cm	1.94	1.28	1.15	3.58	1.58	3.27	0.99	1.66	7.64	4.21	0.00	0.28 #	488
	60 cm	0.06	0.04	0.98	4.97	2.96	0.98	0.01	1.31	4.83	3.10	0.00	0.76 #	191
	90 cm	0.00	0.02	0.99	9.33	3.20	0.13	0.05	1.38	5.50	3.05	0.00	3.77*	193
B.	0 cm(a)	0.35	2.21	1.16	1.73	0.91	0.16	6.04	1.71	4.79	3.83	0.36	1.54 #	
	0 cm(b)	0.35	2.06	1.19	2.87	1.17	0.25	6.01	1.54	6.79	3.69	0.34	1.54 #	551
	10 cm	2.50	1.56	0.94	1.94	0.98	2.52	0.64	1.06	6.01	3.16	0.00	0.83 #	486
	60 cm	0.21	0.38	0.46	0.86	0.99	3.61	0.06	0.83	2.78	2.20	0.00	0.76 #	198
	90 cm	0.20	0.21	0.52	0.73	0.45	3.81	0.02	0.85	1.98	2.98	0.00	0.13 #	206
C.	0 cm(a)	0.80	2.35	1.09	1.96	1.24	0.32	5.40	1.31	6.53	3.12	0.31	2.35 #	
	0 cm(b)	0.80	1.65	1.08	2.41	1.16	0.15	5.22	1.39	5.37	3.05	0.31	2.35 #	554
	10 cm	0.91	1.08	0.91	1.39	0.97	2.92	0.69	1.35	3.31	3.81	0.00	0.40 #	491
	60 cm	0.24	0.30	0.56	0.73	1.00	1.85	0.05	0.92	1.39	2.27	0.00	0.15 #	247
	90 cm	0.14	0.13	0.54	0.80	0.99	2.61	0.05	0.77	1.56	2.25	0.00	0.68 #	247
D.	0 cm(a)	0.00	1.64	1.56	6.63	1.70	0.05	1.00	1.88	2.20	2.79	0.02	5.03$	
	0 cm(b)	0.00	2.20	1.70	5.56	1.47	0.23	7.25	2.39	6.00	4.72	0.27	5.03$	569
	10 cm	0.00	0.64	1.10	17.9	1.57	0.06	0.15	2.06	9.53	4.46	0.00	5.43*	398
	60 cm	0.00	0.01	0.78	10.6	1.88	0.02	0.02	1.32	6.73	3.53	0.00	1.71*	204
	90 cm	0.00	0.02	0.61	13.3	1.16	0.01	0.04	1.20	6.22	2.88	0.00	4.89*	204

mainly organic anions, * mainly bicarbonate, $ about equal organic anions and bicarbonate

the solute fluxes. Postma (1986) and Van Grinsven et al. (subm. for publ.) have evaluated the order of magnitude of errors due to:

(1) spatial variability within a plot,

(2) temporal variability within a (monthly) sampling interval,

(3) errors in the chemical analysis and

(4) uncertainty in the calculated water flux.

The total relative error for the of NO_3^- and Al^{3+} in the acidic soils ranged from 10% (at 10 cm depth) to 30% (at 90 cm depth). Spatial variability within the plots was the dominant source of error.

Table 47 gives the mean annual drainage fluxes of the major solutes at 0, 10, 60 and 90 cm depth in the soil. The fluxes at 0 cm were estimated in two ways. First from mean solute concentrations in forest floor percolates in 1983/84 (see Table 34) and measured water infiltration fluxes (upper row). Second from data on throughfall and litterfall, and various assumptions (lower row). The assumptions were:

1. that all nutrient elements except N are in steady state concentrations in the forest floor,

2. that nutrient uptake by trees is confined to the mineral soil.

From these assumptions it follows that the drainage flux from the forest floor into the mineral soil is equal to the total minus the net uptake plus the atmospheric input. Further assumptions are:

3. that the contribution of nutrient cyling by the ground vegetation is negligible and

4. that the measured H^+ and organic anion concentrations in forest floor percolates apply.

The relative proportions of NO_3^- and NH_4^+ then follow from charge balance considerations.

Except for fluxes of Al^{3+}, N, S and P at plot D, approaches (a) and (b) gave very similar fluxes for most solutes. The similarity may partly be due to the fact that assumptions (1) and (2) above correspond to the situation in the forest floor above the collectors for the 0-cm depth soil solutions, which was unrooted and separated from the undistrubed forest floor by PVC collars (see paragraph 4.3). Lower fluxes for the major nutrients observed in the forest floor percolate collectors at D can be attributed to strongly inhibited mineralization within the collectors. At A, B and C, in contrast to D, a forest floor is actually present in the collectors so litter decomposition would be more similar to that under undisturbed conditions. Especially for nutrients, actual solute fluxes from the forest floor into the mineral soil may have been lower than those reported in Table 47 under b, because of nutrient uptake from the forest floor. The values from the lower row (b) have the greatest and most reliable database, however, and will be used in further considerations.

Anion fluxes at all depths were dominated by NO_3^- and SO_4^{2-}. They were balanced mainly by NH_4^+, K^+, Ca^{2+} and Mg^{2+} just below the forest floor, by Al^{3+} in the acidic mineral soils and by Ca^{2+} in the calcareous soil. Table 47 further shows that:

1. fluxes of H^+, K^+, NH_4^+ and $H_2PO_4^-$ strongly decreased with depth in the upper soil horizons,

2. fluxes of NO_3^-, Na^+ and Cl^- gradually decreased with depth,
3. in the acid soils the fluxes of Al^{3+} increased with depth,
4. high fluxes of Ca^{2+} developed in calcareous horizons and
5. that the fluxes of SO_4^{2-} changed little with depth.

Fluxes of Mg^{2+} increased with depth at plot A, decreased with depth at B, and stayed more or less constant over the soil profile at C and D.

A number of the processes that can explain the patterns discussed above will be discussed next in the framework of H^+ budgets.

6.4 H^+ budgets: the role of atmospheric deposition in soil acidification

Proton budgets of hydrological landscape units summarize the sources and sinks of H^+ in the system considered, and can be used to unravel and quantify the rates of various processes contributing to acidification of soils and waters (Driscoll & Likens, 1982; Van Breemen et al., 1983, 1984). Proton budgets are usually derived from measured or estimated inputs and outputs of ionic solutes in precipitation and drainage water, of dry-deposited gases, and of net assimilation of ionic solutes by vegetation over at least one hydrological year. So, the data presented so far can be used to set up proton budgets. We will not only present H^+-budgets for the whole soil-vegetation systems at each of the four monitoring plots, but also consider H^+-budgets for individual compartments of the ecosystems: tree canopy, forest floor and mineral soil layers of various depths.

To develop H^+-budgets for the different soil horizons, the relative uptake of elements by trees in each of those horizons must be estimated. This was done first by assuming that the nutrient uptake is directly proportional to the transpiration, as calculated during the dynamic simulation of soil water fluxes. The resulting budgets indicated an appreciable accumulation of most cationic nutrients (K^+, Ca^{2+} and Mg^{2+}) in the 0 – 10 cm surface soil, suggesting that the uptake of these nutrients from the surface soil was underestimated. Next we assumed that the difference in the fluxes of K^+ at 0 and at 10 cm depth was due completely to biotic uptake, and that the rest of the K^+ uptake was in the 10-60 cm soil layer. This assumption is reasonable, because the concentrations and fluxes of K^+, certainly in the relatively strongly weathered surface horizons of soils A, B and C, are probably influenced predominantly by biotic uptake, and little by mineral-solution interactions. The somewhat higher uptake of K in the 0 – 10 cm layer indicated by the second model (Table 48) appears realistic given the higher concentrations of nutrients near the soil surface. The high contribution of the 0 – 10 cm surface soil of D to the total uptake of K^+ is in agreement with the intensive shallow root system observed there (Kipp, 1981). All other elements except P and N have been assumed to be taken up according to the 'no-K storage model' in Table 48.

Because the flux of P at 10 cm is very small we have assumed that all uptake of P is confined to the 0 – 10 cm surface soil. In the absence of information on the relative contributions of NH_4^+ and NO_3^- to the total uptake of nitrogen, uptake of N by

Table 48. Distribution of ionic uptake from the soil by trees over the 0-10 cm and 10-60 cm soil layers (% of total uptake), estimated on the basis of two different assumptions, (a) and (b).

	(a) uptake proportional to transpiration				(b) no K^+ storage in the soil at 0-10 cm			
plot:	A	B	C	D	A	B	C	D
depth:								
0-10 cm	18	18	26	37	26	22	30	66
10-60 cm	82	82	74	63	74	78	70	34

plants cannot be considered in the H^+-budget. We followed the procedure Driscoll & Likens (1982), who quantified the net effect of all process involving N-species from the input and output fluxes of ammonium and nitrate. A net input, respectively output, of ammonium to the system constitutes a source, respectively sink, of H^+ of equivalent magnitude. A net input, respectively output, of nitrate constitutes an equivalent sink respectively source of H^+ (Van Breemen et al., 1983). Most of the proton transfers due to N transformations may be due to plant uptake, and nitrogen mineralization plus nitrification, but denitrification and any effects of unseen inputs of gaseous N (N_2, NH_3) are included as well.

In addition to net production of nitrate or net removal of ammonium, other sources of H^+ that have been distinguished are:

1. input of free H^+,
2. net uptake of cations by biota and/or net mineralization (or leaching from the tree canopy) of anions other than N-species,
3. net accumulation of cations (excluding Al^{3+}) or net removal of anions from the mineral soil and
4. net accumulation of Al^{3+} in the mineral soil.

Similarly, in addition to H^+ – consuming processes involving NH_4^+, sinks of H^+ include:

1. output of free H^+,
2. net mineralization (or canopy leaching) of cations and/or net uptake of anions other than N-species,
3. net removal of cations (excluding Al^{3+}) or net accumulation of anions in the mineral soil and
4. net removal of mineral Al^{3+}.

These components of the H^+-budget have been quantified from the data of Tables 41, 46 and 47, assuming no changes in storage of dissolved ions over a hydrological year in the compartments considered. The net flux of a solute into or from a mineral soil compartment, calculated from solute transport across the compartment boundaries (0, 10, 60 and 90 cm depth) and ionic uptake from the compartment, was considered to be due to dissolution or precipitation of solid phases plus ion exchange.

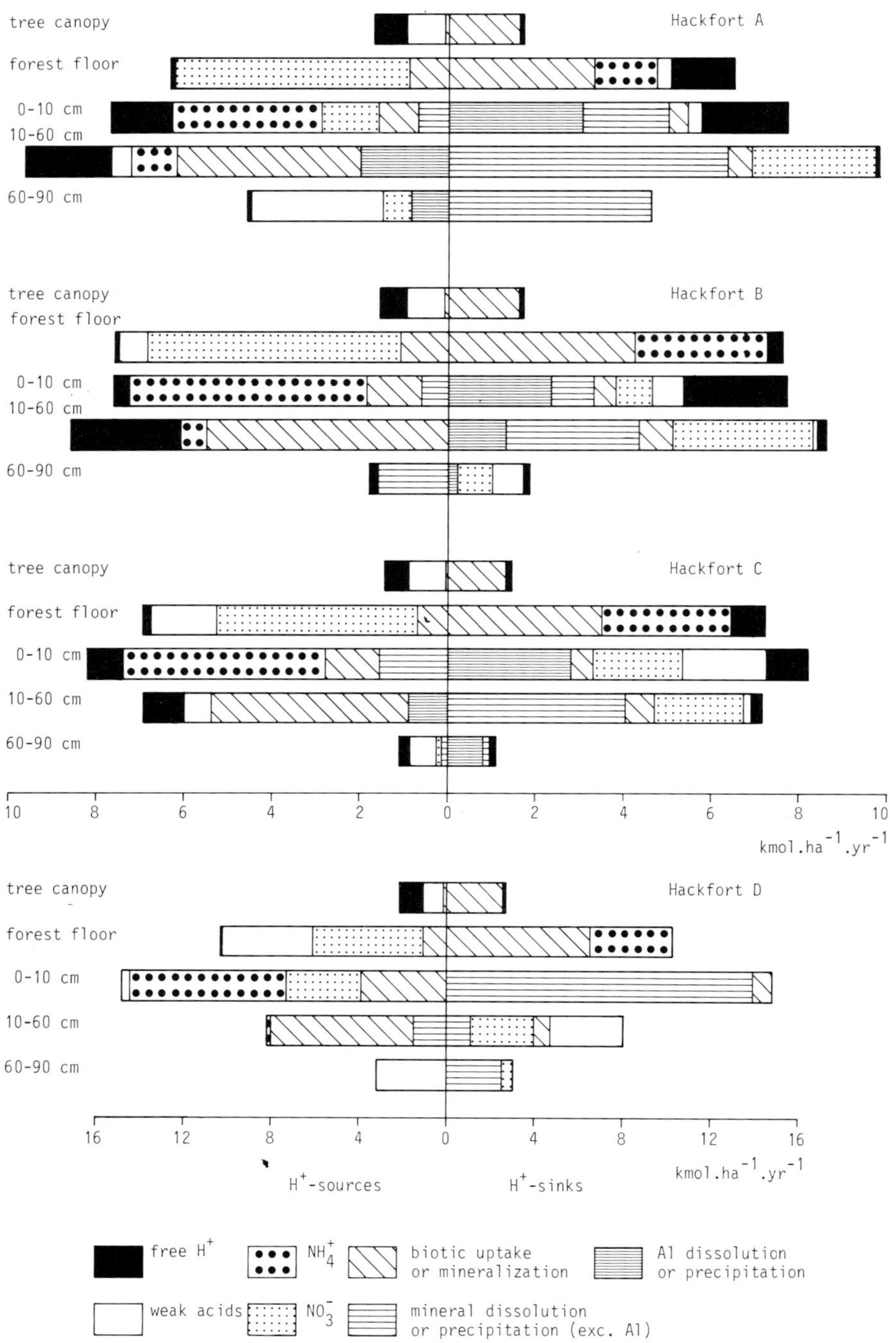

Figure 34. Proton budgets for the tree canopy, the forest floor, and three layers of the mineral soil (0 – 10, 10 – 60, and 60 – 90 cm depth) at each of the plots A, B, C and D. Values are simple means for three hydrological years.

The resulting budgets for the compartments tree canopy, forest floor and mineral soil, subdivided into two rootzone compartments (0 – 10 cm and 10 – 60 cm) and the non-rooted subsoil (60 – 90 cm), have been summarized in Figure 34. The budgets for specific compartments in different plots are generally rather similar from plot to plot, and will be discussed per compartment.

Tree canopy

In the tree canopy, the two dominant sources of H^+ are about equally important: external H^+ inputs (mainly due to oxidation of SO_2 and NO_x), and dissociation of weak acids (mainly organic acids produced by the trees, with minor amounts of CO_2). These sources are almost completely (> 90%) balanced by leaching of base cations, so that little free H^+ remains in throughfall water. The total H^+-transfer in tree canopies is considerably higher at the calcareous plot D than at the plots with acidic soils (note, in Fig. 34, the difference in scale between A, B and C versus D). The difference is due to higher atmospheric inputs, and a larger contribution of weak acid dissociation, with a concomitant increase of leaching of base cations (mainly Ca^{2+}) from the canopy.

Forest floor

Nitric acid (from nitrified organic N and atmospheric NH_4^+) and organic acids are the main H^+ sources in the forest floor, with a relatively small contribution of phosphoric and sulfuric acids from mineralized organic P and S. In the forest floor at plot A organic acids do not dissociate but, on the contrary, part of the organic anions in water leaching from the canopy, protonate and so serve as H^+ sinks. Depressed solubility and H^+ dissociation due to high nitrification in the forest floor may cause this phenomenon.

Most of the H^+ formed in the forest floor is neutralized by formation of ammonium and by mineralization of base cations. However, some free H^+ remains and, contrary to the tree canopy compartment, the forest floor is a source of free H^+, at least in case of acid soils: H^+ production amounts to 1.3 (A), 0.2 (B) and 0.7 (C) kmol/ha.yr. The calcareous soil D has, in fact, no forest floor because essentially all tree litter is decomposed within a year. Litter decomposition there (the 'forest floor') results in a further neutralization of the little free H^+ that remained in the throughfall.

Mineral soil

The upper 10 cm of the mineral soil at acidic sites again generates free H^+, viz. 0.6(A), 2.1.(B) and 0.1(C) kmol/ha.yr. Main H^+ sources in this layer are uptake of ammonium, and, at A and D, nitrification of part of the ammonium. Partial neutralization takes place through protonation of organic anions percolating from the forest floor, and by dissolution of aluminium (A, B and C) and calcium (D) from soil minerals. In the surface soil of plots B and C, no net nitrification takes place. In fact, part of the nitrate percolating from the forest floor is removed, presumably by plant uptake. The highest rates of acid production are observed in the calcareous surface soil (20 kmol/ha.yr) with removal of NH_4^+, formation of nitrate and uptake of base cations as the dominant H^+ sources. This H^+ is completely neutralized, mainly by dissolution of calcium carbonate.

The acid mineral soil at 10 – 60 cm neutralizes most of the free acid coming from the surface layer. Removal of nitrate (presumably mainly uptake by plants, perhaps partly by denitrification) and weathering of base cations are the main sinks of H^+. Another sink is removal of sulfate, which accounts for the buffering of 0.6(A), 0,4(B), 1.0(C) and 0.5(D) kmol/ha.yr of H^+. Removal of sulfate will be due to precipitation of basic alumininium sulfate (at least in A and C, where Al^{3+} also precipitates in this horizon (Mulder et al., subm. for publ.)), microbial transformation to organic sulphur (David et al., 1982), or adsorption of SO_4^{2-} on Fe(III)-oxides (see par. 2.4). In the calcareous 10 – 60 cm subsoil of plot D, part of the dissolved calcium bicarbonate percolating from the surface soil precipitates so that here HCO_3^- serves as a H^+ sink. Below the rootzone, low rates of H^+ transfer are observed in the acid soils B and C. In the calcarous subsoil some calcium carbonate dissolution takes place in this horizon.

H^+-budgets of the whole soil-vegetation systems

The role of atmospheric deposition in the H^+ budget of the overall budgets can be obtained by summing the budgets for individual compartments shown in Fig. 34, cancelling opposing processes in different compartments (e.g. nitrification at 0 – 10 cm depth and net nitrate removal at 10 – 60 cm will result either in net nitrification or net nitrate uptake at 0 – 60 cm depending on which is the dominant process). They can also be calculated directly from atmospheric input, net uptake by trees and drainage fluxes at 90 cm depth. The H^+ sources have been divided into two categories, internal and external. The external H^+ sources are dry plus wet deposited H^+ (H_t in Table 41), and atmospheric ammonium. Because essentially no ammonium is removed from the system by drainage, and because part (at B and C) or all (at A and D) of the NH_4^+-N input is recovered as NO_3 in the drainage at 90 cm depth, between 1 and 2 kmol of H^+ are produced per kmol of NH_4^+ deposited from the atmosphere:

$$NH_4^+ \rightarrow \text{organic N} + H^+$$

$$NH_4^+ \rightarrow NO_3^- + 2H^+$$

Table 49 shows that uptake and nitrification of atmospheric ammonium is the main H^+-source (3.0 to 6.8 kmol/ha.yr), exceeding the total free H^+ input by a fac-

Table 49. H^+-budgets for the four soils A, B, C and D. Values are means for the April 1981 to 1984 period, in kmol/ha.yr.

	H^+ sources								H^+ sinks internal				
	external		internal										
	H^+	N	N	CO_2 org.acid	biom.	wth.#	sum		sum	weathering Al^{3+}	weathering other*	outp.	ANC
plot:													
A	0.7	5.8	1.6	3.8	0.2	0.0	12.1		12.4	0.1	12.3	0.0	-12.3
B	0.6	3.9	0.0	0.0	0.8	0.6	5.9		6.0	3.8	2.0	0.2	- 5.3
C	0.4	3.0	0.0	0.7	0.9	0.4	5.4		5.4	2.6	2.7	0.1	- 4.9
D	1.1	6.8	1.6	4.9	1.3	0.9	16.7		17.2	0.0	17.2	0.0	-16.3

anion dissolution plus cation precipitation
* cation dissolution (excl. Al^{3+}) plus anion precipitation

tor of six to eight. In plots A and D even more H^+ is generated as a result of nitrogen transformations than can be accounted for from external inputs. The extra H^+ formation must be ascribed to mineralization and nitrification of organic N, an internal source. At D the extra N may come from fixation of atmospheric N_2 by alder (Van Miegroet et al., 1984). At A, trees have grown very poorly in the past five years (de Visser, 1986; see also the data on net uptake in Table 8), and decomposition may now exceed accumulation of soil organic N (Van Breemen et al., subm. for publ.). Uptake of cations by biomass and, even in calcareous soil, dissociation of CO_2 are minor sources of H^+.

The only sink of H^+ that is quantitatively important when considering the whole soil + vegetation system is mineral weathering. Table 50 shows the changes in the pools of individual components due to mineral weathering. In soils A and D, calcium carbonate weathering is the main sink for H^+. In soils B and C dissolution of aluminium is predominant (mainly in the surface soil), with additional dissolution of base cations (mainly in the subsoil). At A too, strong dissolution of aluminium

Table 50. Annual changes in the pools of various components in the mineral soil (0-90 cm) at each of the four plots. Means for three hydrological years (1981-1984), in equivalent kmol/ha.yr.

	Component							
plot:	K_2O	Na_2O	CaO	MgO	Al_2O_3	HCl	SO_3	P_2O_5
A	-0.04	0.03	- 9.00	-3.02	-0.13	0.05	0.12	0.00
B	-0.41	0.53	- 0.71	-0.36	-3.81	0.55	0.01	-0.02
C	-0.37	0.40	- 0.82	-0.90	-2.60	0.48	0.15	-0.02
D	-0.20	0.88	-13.90	-1.03	-0.01	0.97	1.12	-0.03

takes place in the surface horizons, but practically all of it is precipitated in and just above the calcareous subsoil, so that net aluminium weathering over the 90 cm deep soil profile is almost nil. A striking aspect of mineral weathering at A is the large contribution of Mg. The large storage of NaCl (Na_2O + HCl) at plots B, C and D, indicated by Table 50 is probably not realistic, and may be due to a combination of

1. overestimated atmospheric inputs,
2. overestimated drainage rates below the rootzone and
3. biased sampling with respect to spatial variability of soil solution concentrations at 90 cm depth.

The net decrease in the pools of cationic components plus the net increase in the pools of anionic components of Table 50 is equal to ANC, the net decrease in acid neutralizing capacity of the soil. ANC, shown in Table 49, can be considered as a measure of the rate of soil acidification (Van Breemen et al., 1983). For the calcareous soil, the value of ANC can be considered normal, although the nature of the dominant H^+ source ($NH_2^+ \rightarrow NO_3^- + 2H^+$) is clearly influenced by human action. Under natural conditions CO_2 can be an equally effective H^+ source at the near neutral pH provided by buffering of calcium carbonate. In the acid soils, the rate of decrease in ANC (4 – 5 kmol/ha.yr) is two to ten times higher than the rate that can be expected under naturel forest conditions on acidic soils (Mazzarino et al., 1983; Breeuwsma & de Vries, 1984). Clearly, under the influence of acid atmospheric deposition the total H^+ load to the forest soils has increased considerably, resulting in increased mineral weathering and a shift to H^+ buffering by aluminium.

7 Summary and conclusions

This part describes the results of three years (1981 – 1984) of biogeochemical monitoring of an oak woodland in the Netherlands, including the methodology applied.

7.1 Description of the research

The study area is a small 3.2 ha old coppice woodland (last cut in 1939, in an area with about 75% grassland for dairy cattle, and about 25% forested land, wooded strips and hedges. The soils have developed from mineralogically rich sandy to loamy, Pleistocene Rhine sediments, originally calcareous but now largely decalcified. Four plots, about 10 × 20 m^2, with different soils were selected for biogeochemical monitoring:

A. an acid (pH 3.5) brown loamy soil with a calcareous subsoil starting at 80 cm below the soil surface,

B. an acid (pH 3.5) brown sandy soil with a calcareous subsoil below 120 cm,

C. an moderately acid (pH 4), yellowish brown sandy soil without calcareous material and

D. a grey, sandy soil that is calcareous throughout.

The three acidic soils have an organic surface horizon ('forest floor') of 3 to 6 cm thickness. The groundwater table is generally within 150 cm (B) to 120 cm (A, C and D) below the land surface, but may be within the rootzone (0 – 50 cm depth) for several days or weeks during winter and early spring.

The tree vegetation is dominated by oak (*Q. robur*) at B and C, and consists of a mixture of oak and birch (*B. pendula*) at A, and of oak, birch, alder (*A. glutinosa*) and poplar (*P. tremula*) at D. The ground vegetation is rather rich at all sites, but particularly at the calcareous plot D. In summer bracken (*Pteridium aquilinum*) is the dominant ground cover at the plots with acidic soils.

The monitoring programme consisted of regular sampling and chemical analysis of rain and snow water, throughfall (or canopy drip), stemflow (or trunkflow) and soil solutions at various depths, collected by suction from permanently installed ceramic porous cups. Soil water movement was estimated by means of a simulation programme using climatic and soil physical input data, and measured hydraulic potentials as fitting parameters. Solute fluxes were quantified by multiplying monthly measured (above-ground) or simulated (below-ground) water fluxes with measured concentrations.

In addition, tree litter was collected and analysed for two years, and surveys were made of the ground vegetation and of the fungus flora. Results of these surveys were compared with those carried out in the fifties and sixties in the same area. Similar comparisons were made for data on the forest floor (Winkels, 1985). Inventories of tree biomass, and reconstructions of the growth of sampled trees and of the stands at the plots were done in a study reported elsewhere (De Visser, 1986).

7.2 Results of the hydrological and chemical monitoring

The three years during which monitoring took place were rather similar in hydrological characteristics, with annual precipitation close to the long-term mean of 700 to 800 mm. Throughfall was close to 80% of the precipitation in the open field, both in summer and winter. The contribution of stemflow to water supplied to the soil in the woodland was 1 – 2% of total precipitation. Precipitation water collected underneath the canopy was strongly concentrated in most solutes. Of all solutes, ammonium and sulfate showed the highest fluxes, which were increased by a factor 2 – 4 on passage through the canopy. Annual inputs of N and S into the soil from throughfall and stemflow were 2 – 4 (equivalent) kmol per ha, with highest values at plot D and lowest at plot C.

Apart from organic carbon, which occurred in high concentrations (2 – 12 mmol/l), particularly at shallow depth, the soil solutions were dominated by calcium (annual mean 5 – 6 equivalent mmol/l) and nitrate (2 – 3 mmol/l) plus sulfate (1 – 1.5 eq. mmol/l) at the calcareous plot D, and by aluminium (0.5 – 2 eq. mmol/l), nitrate (0.7 – 2.5 mmol/l) and sulfate (0.5 – 1.5 eq. mmol/l) in the acidic soils of the plots A, B and C. In the least acid soil C, concentrations of these ions were distincly lower than in A and B. The nitrate concentrations are very high compared to published data on other forest soils. Except at plot D, where atmospherically fixed N from alder may contribute to this phenomenon, the high nitrate concentrations indicate:

1. that ammonium nitrogen from atmospheric inputs is nitrified, even at very low soil pH and
2. that supply of N exceeds the demand by the vegetation.

Nitrate concentrations are strongly seasonal, at plots A, B and D characterized by an apparent build-up of near the soil surface through summer and fall, and a downward movement of the zone with high nitrate through winter and spring. At plot C, on the other hand, lowest nitrate concentrations are observed in the root zone during summer, indicating that a larger fraction of the available nitrogen is used by the vegetation than at the other plots.

Nitrification and the formation of nitric acid associated with that process leads to permanent soil acidification if the nitrate is not assimilated by plants and thus has the opportunity to carry away cations in the drainage water. Nitrification of atmospherically derived ammonium is indeed the dominant soil acidification process in all plots. The acidification is particularly strong in the 0 – 10 cm surface horizon, but is partly reversed by nitrate removal (presumably by plant uptake) in the lower part of the root zone. Acidification is associated with dissolution of calcium carbonate in the calcareous soils and of mineral or adsorbed aluminium in the acidic soils. The high aluminium concentrations in the soil solution at A and B can be attributed mainly to dissolution of aluminium by nitric acid.

Unfortunately, insufficient 'old' (1956-1960) soil chemical data were available to evaluate the magnitude of possible soil chemical changes. Comparison of old and new data by Winkels (1985) indicated that the forest floor has increased strongly: C pools have grown by factors of 1.25 (B), 2.2 (C) and 2.7 (A). Moreover, the C:N ratio has decreased from 28 to 23. Whereas the increase in the forest floor is probably a normal process in the development of the stand, the decrease in C:N ratio is probably due to increased N availablity from atmospheric deposition.

7.3 Tree growth and changes in vegetation

The ground vegetation had changed markedly over the past decades, with a strong expansion of species favoured by a high supply of nitrogen. Changes in ground vegetation do not indicate that there have been marked changes in soil acidity, in soil moisture availability and in light intensity. Because the strong soil acidity associated with high concentrations of dissolved aluminium nitrate has most probably developed over the past decades as a result of strongly increased atmospheric ammonium inputs, it may appear surprising that such a change in soil chemical conditions is not reflected by the composition of the ground vegetation. A possible explanation is that at least some species of the ground vegetation may root mainly in the organic forest floor, where aluminium concentrations are very low.

The fungus flora has changed in the past decades, with a relative decline of mycorrhizal fungi and a relative increase in the occurrence of saprophytes. Decline of mycorrhizal fungi may be related to increased ammonium supply, either directly or due to increased soil acidity. An increase in saprophytes may be related to the increased thickness of the forest floor, and accumulated dead wood, both of may be due mainly to normal processes in the development of a relatively young, even-aged, forest stand.

From the study by De Visser (1986), part of which was summarized very briefly in paragraph 3.1., the tree vegetation at plot A appears to have grown very poorly over the past 20 years. Plots B and C, on the other hand, have grown well in comparison with stands described in the literature. Also, at plot A leaf- and branch-litter production is distinctly lower than at the other plots. Poorer growth at A than at B and C cannot be explained simply by aluminium toxicity: concentrations of dissolved inorganic aluminium in the root zone are about equally high at A (poor growth) and B (good growth) and distinctly lower at C (good growth). Molar ratios of dissolved Ca^{2+}/Al^{3+} in the root zone, which may be a better indicator of aluminium toxicity than Al^{3+} concentrations alone, are about equally low (generally 0.5 – 1) at all three plots. However, plot A differs from the other plots in having distinctly higher levels of manganese, in the soil solution as well as in canopy drip and in leaf litter. The possibility that the high Mn^{2+} concentrations in the soil solution are related to the growth decline at plot A (personal communication, M. Kazda) merits further investigation. The high manganese concentrations could be a direct result of increased soil acidification by atmospheric deposition. Higher manganese concentrations in soil solutions at plot A than at plot B (both of which have about the same pH) can be explained by higher Mn^{2+} contents in soil material (and presumably better availability of Mn^{2+} for buffering protons) at A than at B.

7.4 Nitrogen dynamics and soil acidification

That nitrification can take place in highly acid forest soils has been described by many others (see for a short review Van Breemen et al., subm. for publ.). However, extreme soil acidification by nitrification, as described here, is rare. The reason for the strong acidification by nitrification must be ascribed to the fact that the biota (i.e. plants plus soil micro-organisms) are not able to utilize all the nitrogen available. That nitrogen is present in excess can be due to a combination of two factors: high rates of supply (both from the atmosphere and from mineralized N) and low rates of uptake. Judged from tree growth the rate of uptake (both by trees and by micro-organisms, which at least in part depend on short-term supply of energy by higher plants) is presumably low at A and distinctly higher at C. This is corroborated by higher nitrate levels and higher drainage fluxes at A than at C. As indicated by Table 51, drainage of N (essentially only nitrate) even exceeds atmospheric N inputs (mainly as ammonium) at A, implying that the pool of soil organic nitrogen serves as a source of nitric acid. At plot B, tree growth is slightly better than at C (De Visser, 1968), in accordance with a higher retention of N. However, both inputs and drainage rates of N are higher at B than at C. At D, N_2 fixation by Frankia in symbiosis with alder may have lead to an underestimation of the input, and thus account for a (possibly apparent) negative retention of nitrogen.

If differences in vitality of the trees are indeed a major reason for differences in N

Table 51. Annual input (in throughfall) and output (in drainage water at 90 cm depth) of inorganic N, and the inorganic N balance in the soil between 0 and 90 cm depth (means ± s.d. for three hydrologhical years, 1981-1984). From Van Breemen et al., subm. for publ.

	N-flux (kg/ha.yr)		
plot:	throughfall input	drainage output	retained
A	54.6 ± 4.6	78.5 ± 18.6	-24.0 ± 23.4
B	56.2 ± 5.1	28.1 ± 12.4	28.1 ± 11.6
C	44.6 ± 2.8	22.5 ± 10.7	22.1 ± 9.7
D	62.8 ± 6.8	87.6 ± 26.5	-25.1 ± 21.8

dynamics, it could be generally true that decreased tree vitality may lead to decreased uptake of N, which in turn can lead to a 'saturation' of the ecosystem with N, resulting in increased soil acidification, partly by net nitrification and drainage of N from the soil organic matter pool (Van Breemen et al., subm. for publ.). Of course, the higher the rate of N input, the sooner a stage of 'N-saturation' would be reached. In terms of atmospheric N supply, the situation at the 'Oude Maat' is extreme by global or western European standards. However, a similar trend towards 'N-saturation' may be present in forests in e.g. the Federal Republic of Germany, where N inputs are lower, but the decrease in tree vitality is sometimes more serious.

8 References

Adema, E.H., P. Heeres, & J. Hulskotte, 1985. On the dry deposition of NH_3, SO_2 and NO_2 on wet surfaces in a small scale windtunnel. Paper presented during the 7th World Clean Air Congress, Sydney, 25-29 September 1986: 8 pp.

Altena, B., 1985. De bepaling van verschillende zwavelfracties in de bodem. Unpublished Report, Dept. of Soil Science and Geology, Agricultural University, Wageningen, The Netherlands.

Arnolds, E., 1984. Standaardlijst van Nederlandse macrofungi. Coolia 26, suppl.

Arnolds, E. (ed.), 1985. Veranderingen in de paddestoelenflora (mycoflora). Wetenschappelijke Mededeling 167. Kon. Ned. Natuurhistorische Vereniging.

Arts, H., 1986. De beschrijving van een model dat de mogelijkheid biedt de hoeveelheden nat en droog gedeponeerd materiaal rekenkundig te onderscheiden van de hoeveelheden van dat materiaal die door de plant (netto) zijn opgenomen of afgestaan. Unpublished M.Sc. Thesis. Depts. of Air Pollution and of Soil Sci. and Geology, Agric. Univ. Wageningen, the Netherlands, 41 p. + 76 pp.

Barkman, J.J., H. Doing & S. Segal, 1964. Kritische Bemerkungen und Vorschläge zur Quantitativen Vegetationsanalyse. Acta Bot. Neerl. 13: 394-420.

Bascomb, C.L., 1964. Rapid method for the determination of cation exchange capacity of calcareous and non-calcareous soils. J. Sci. Food. Agric. 15: 821-823.

Begheijn, L.Th., 1980. Methods of chemical analyses for soils and waters. Internal Report, Dept. of Soil Science and Geology, Agric. Univ. Wageningen, The Netherlands.

Begheijn, L.Th., 1981. Preparation and use of standards for chemical analysis of water samples. Internal Report, Dept. of Soil Science and Geology, Agric. Univ. Wageningen, The Netherlands.

Belmans, C., J.G. Wesseling & R.A. Feddes, 1981. Simulation model of the water balance of a cropped soil providing different types of boundary conditions. ICW Note 1257. Wageningen, The Netherlands.

Belmans, C., J.G. Wesseling & R.A. Feddes, 1983. Simulation model of the water balance of a cropped soil: SWATRE. J. of Hydrology 63: 271-286.

Black, T.A., W.R. Gardner & G.W. Thurtell, 1969. The prediction of evaporation, drainage and soil water storage for a bare soil. Soil Sci. Soc. Amer. Proc. 33: 655-660.

Bloemen, G.W., 1980. Calculation of hydraulic conductivities of soils from texture and organic matter content. Z. Pflanzenernähr. Bodenk. 143: 581-605.

Breeuwsma, A. & W. de Vries, 1984. The relative importance of natural production of H+ in soil acidification. Neth. J. Agric. Sci. 32: 161-163.

Brewer, R., 1964. Fabric and mineral analysis of soils. John Wiley & Sons Inc. N.Y., London, Sydney.

Buddingh, C. & H. Vissers, 1981. Kartering van een boscomplex in de omgeving van Hackfort. Unpublished M.Sc. thesis. Dept. of Soil Science and Geology, Agric. Univ. Wageningen, The Netherlands, 48 pp.

Buijsman, E., H. Maas, & W. Asman, 1984. Een gedetailleerde ammoniakemissiekaart van Nederland. Instituut voor Meteorologie en Oceanografie, University of Utrecht, The Netherlands. Rapp. V-84-20: 124 pp.

Dane, J.H., & S. Hruska, 1984. In-situ determination of soil hydraulic properties during drainage. Soil Sci. Soc. Amer. J. 47: 619-623.

Davis, M.B., M.J. Mitchell & J.P. Nakas, 1982. Organic and inorganic sulfur constituents of a forest soil and their relationship to microbial activity. Soil Sci. Soc. Amer. J. 46: 847-852.

De Bakker, H. & J. Schelling, 1966. Systeem van bodemclassificatie voor Nederland. Pudoc, Wageningen, 217 pp.

Dekker, C.G., 1979. KNMI, Verslagen V-137. Een onderzoek naar de grootte van de systematische windfout van de standaard regenmeter. Royal Meteorological Institute, De Bilt, the Netherlands, Rept. V-137.

De Laat, P.J.M., 1980. Model for unsaturated flow above a shallow watertable, applied to a regional subsurface flow problem. Agric. Res. Rep. 895, PUDOC, Wageningen, 126 pp.

Distel, A.C. & A. van Straten, 1980. De bodemgesteldheid rond het bos 'de Oude Maat' bij Hackfort in de gemeente Warnsveld. Unpubl. M.Sc. thesis, Dept. of Soil Science and Geology, Agric. Univ. Wageningen, The Netherlands, 78 pp.

Dobben, H. van & J. Mulder (eds) (In prep.). Biogeochemistry of moorland pools and surrounding uplands in the Netherlands, affected by acid atmospheric deposition. Agric. Res. Rept. PUDOC, Wageningen, The Netherlands. In prep.

Driscoll, C.T., & G.E. Likens, 1982. Hydrogen ion budget of an aggrading forested exosystem. Tellus 34: 283-292.

Ellen, H. Zeigerwerte der Gefasspflanzen Mitteleuropas. Verlag Erich Goltze KG, Göttingen, 1974.

FAO, 1977. Guidelines for soil profile description. FAO, Rome, 66 pp.

Feddes, R.A., P.J. Kowalik, & H. Zaradny, 1978. Simulation of field water use and crop yield. Simulation Monographs. Pudoc. Wageningen, 48 pp.

Heymans, J.M.G.B. & H. Tuinhof, 1983. Sedimentatie en bodemkundige inzichten nader bekeken met gebruik van statistische verwerkingsmethoden op een raai nabij Hackfort. Unpubl. M.Sc. thesis, Dept. of Soil Science and Geology, Agric. Univ. Wageningen, The Netherlands, 75 pp.

Hillel, D., 1980. Fundamentals of soil physics. Academic Press, New York, pp. 125-155.

Hoeksema, K.J. & C.H. Edelman, 1960. The role of biological homogenisation in the formation and transformation of grey-brown podzolic soils. Trans. 7th. Int. Congress Soil Sci., Madison, Vol IV, pp. 402-405.

Houba, V.J.G., I. Novozamsky, J.J. van der Lee, W. van Vark & E. Nab, 1985. Chemische analyse van gewassen. Department of Soil Science and Plant Nutrition, Agric. University Wageningen, the Netherlands, 62 pp.

Jansen, A.E., 1981. The vegetation and macrofungi of acid oakwoods in the north east Netherlands. Thesis, Wageningen. Handelseditie 1984, Agricultural Research Reports 923, PUDOC, Wageningen.

Jong, de W.W., 1949. Geografische aantekeningen uit de gemeenten Steenderen en Bronkhorst. Tijdschr. Kon. Ned. Aardr. Genootsch. 66: 1-34.

Jongmans, A.G., 1980. Verslag van het practicum Verband Bodem en Vegetatie F 250-02/J 100-034; Hoofdstuk III, veldbodemkunde, pp. 23-40; Dept. of Soil Science and Plant Nutrition, Agric. Univ. Wageningen, The Netherlands.

Jongmans, A.G., 1981. Vakgroepexcursie 1981, Hackfort. Dept. of Soil Science and Geology, Agric. Univ. Wageningen, The Netherlands. 10 pp.

Jongerius, A. & G. Heintzberger, 1963. The preparation of mammoth-sized thin sections. Soil Survey Papers 1, Soil Survey Inst., Wageningen, the Netherlands, 37 pp.

Kipp, J.A., 1981. Bewortelingsonderzoek ten behoeve van het Hackfort projekt (Research on root distribution for the Hackfort project). Unpubl. M.Sc. Thesis. Dept. of Soil Science and Geology, Agric. Univ. Wageningen, The Netherlands.

Klute, A., 1972. The determination of the hydraulic conductivity and the diffusivity of unsaturated soils. Soil Sci. 113: 264-276.

Knibbe, M., 1969. Gleygronden in het dekzandgebied van Salland. Communications of the Soil Survey Institute. Bodemk. Studies 8, Wageningen, the Netherlands.

Kools, J.P. & A. Ehrenburg, 1985. Een vergelijkend vegetatieonderzoek in een eikenhakhoutbos bij Hackfort (1957-1984), Unpublished M.Sc. Thesis. Vegetation Science, Plant ecology and Weed science, Agricultural university Wageningen, the Netherlands.

Mazzarino, M.J., H. Heinrichs & H. Fölster, 1983. Holocene versus accelerated actual proton consumption in German forest soils. In: Effects of accumulation of air pollutants in forest ecosystems. Ed. by B. Ulrich and J. Pankrath: 113-123, Reidel Publ. Co. Dordrecht, Holland.

Mexal, J. & C.P.P. Reid, 1973. The growth of selected mycorrhizal fungi in response to induced water stress. Can. J. Bopt. 51: 1579-1588.

Miedema, R., E. Van Engelen & Th. Pape, 1978. Micromorphology of a toposequence of Late Pleistocene fluviatile soils in the Eastern part of the Netherlands. In: Delgado (ed.): Micromorphologia de Suelos. Universidad de Granada, Vol. 1, p. 469-501.

Minderman, G. (ed.), 1981. Bodemkundige en bosecologische gegevens van het eikenhakhoutbosje Hackfort in de gemeente Vorden. Part I to IV. Research Institute for Nature Management, Arnhem, The Netherlands.

Mitscherlich, G., 1971. Wald, Wachstum und Umwelt. T. II. Frankfurt a. M., Sauerländer's Verlag, 365 pp.

Moser, M., 1983. Die Rörlinge und Blätterpilze (Polyporales, Boletales, Agraricales, Russulales), 5th ed. Fischer. Stuttgart.

Mulder, J.P.M., 1983. A simulation of rainfall interception in a pine forest. PhD. Thesis, State University Groningen. 51 pp.

Mulder, J., J.J.M. van Grinsven & N. van Breemen, 1987. Impact of acid atmospheric deposition on woodland soils in the Netherlands. III, Aluminum chemistry. Subm. for publ. Soil Sci. Soc. Amer. J.

North American Conference on Mycorrhizae, 1981. Program and abstracts., Quebec, Canada.

Oliver, B.G., E.H. Thurman & R.L. Malcolm, 1983. The contribution of humic substances to the acidity of colored natural waters. Geochim. Cosmoch. Acta 47: 2031-2035.

Onderdelinden, D., J.A. van Jaarsveld & N.D. van Egmond, 1984. Bepaling van de depositie van zwavelverbindingen in Nederland. Nat. Inst. Public Health and Env. Hygiene, Bilthoven, the Netherlands, Rept. 842017001, 39 pp.

Pannekoek, A.J., 1973. Algemene Geologie. Tjeenk Willink, Groningen, the Netherlands. 533 pp.

Pape, J.C., 1972. Oude bouwlandgronden in Nederland. Boor en Spade 18, Veenman en Zn., Wageningen.

Parker, G.G., 1983. Throughfall and stemflow in the forest nutrient cycle. In: Adv. Ecol. Res., Vol. 13. Ed. by McFadyer and Ford. Academic Press: 58-120.

Postma, H., 1986. Foutenberekening bij het bodemonderzoeksproject in Hackfort. Unpubl. M.Sc. thesis, Department of Soil Science and Geology, Agricultural University, Wageningen, the Netherlands.

Roberts, J., 1983. Forest transpiration: a conservative hydrological process. J. of Hydrology 66: 133-141.

Roelofs, J.G.M., A.J. Kempers, A.L.F.M. Houdijk, & J. Jansen, 1985. The effect of air-borne ammonium sulphate on Pinus nigra var. maritima in the Netherlands. Plant and Soil 84: 45-56.

Rutter, A.J., 1968. Water consumption by forests. In: T.T. Kozlowski (ed.). Water deficit and plant growth II, p. 23-84. Acad. Press, New York.

Schroeder, M.J. & C.C. Buck, 1978. Fireweather U.S. Department of Agriculture Forest Service. Agricultural Handbook 360.

Singh, B. and G. Szeicz, 1979. The effect of intercepted rainfall on the waterbalance of a hardwood forest. Water Resources Res. 15: 131-138.

Stiboka, 1979. Bodemkaart van Nederland 1:50,000. Blad 33 W/O, Apeldoorn. Soil Survey Institute, Wageningen, The Netherlands.

Ulrich, B., 1983. A concept of forest ecosystem stability and of acid deposition as driving force for destabilization. In: Effects accumulation of air pollutants in forest ecosystems. Ed. by B. Ulrich and J. Pankrath: 1-29, Reidel Publ. Co., Dordrecht, Holland.

USDA, Soil Conservation Service, 1975. Soil Taxonomy. U.S. Government Printing Office, Washington D.C., 754 pp.

Van Aalst, R.M., H.S.M.A. Diederen, J.C.T. Hollander, K.D. van der Hout, F.A.A.M. de Leeuw & J. Stellingwerf, 1983. Gevolgen van emissies van stikstofoxiden in de lucht: volgproducten van NO_x, fotochemische luchtverontreiniging en zichtvermindering in Nederland. TNO, Delft, the Netherlands. Rapp. R83/58: 42 pp., 6 bijl.

Van Aalst, R.M. & H.S.M.A. Diederen, 1982. De rol van stikstofoxiden en ammoniak bij de depositie vanuit de lucht van bemestende en verzurende stoffen op de Nederlandse bodem. Organisatie voor Toegepast Natuurwetenschappelijk Onderzoek, TNO, Hoofdgroep Maatschappelijke Technologie, the Netherlands. R 83/42: 30 pp.

Van Bakel, P.J.T. & T.J. van de Nes, 1984. Herziening van de berekening van de gewas verdamping in het hydrologische model GELGAM. Rapp. Adhoc groep verdamping, Rijkswaterstaat, Arnhem, the Netherlands, 92 pp.

Van Breemen, N., P.A. Burrough, E.J. Velthorst, H.F. van Dobben, T. de Wit, T.B. Ridder & H.F.R. Reijnders, 1982. Soil acidification from atmospheric ammonium sulphate in forest canopy throughfall. Nature, 299: 548-550.

Van Breemen, N. & E.R. Jordens, 1983. Effects of atmospheric ammonium sulfate on calcareous and non-calcareous soils of woodlands in the Netherlands. In: B. Ulrich and J. Pankrath (eds.), Effects of Accumulation of Air Pollutants in Forest Ecosystems, p. 171-182. Reidel Publishing Company, 1983.

Van Breemen, M., C.T. Driscoll & J. Mulder, 1984. Acid deposition and internal proton sources in acidification of soils and waters. Nature 307: 599-604.

Van Breemen, N., P.H.B. de Visser & J.J.M. van Grinsven, 1986. Nutrient and proton budgets in four soil-vegetation systems underlain by Pleistocene alluvial deposits. J. Geol. Soc. London, 143: 659-666.

Van Breemen, N., J. Mulder & J.J.M. van Grinsven, 1987. Impact of acid atmospheric deposition on woodland soils in the Netherlands, II. N-transformations. Subm. for publ., Soil Sci. Sco. Amer. J.

Van de Meene, E.A., 1979. Het ontstaan van de Gelderse IJssel, KNAG. Geografisch Tijdschrift XIII, 202-210.

Van Dis, A., 1984. Micromorfologische beschrijving van vier bodemprofielen in 'de Oude Maat' (Hackfort). Internal Report, nr. 52, section morphology, Dept. of Soil Science and Geology, Agric. Univ. Wageningen, The Netherlands.

Van Genuchten, M.Th., 1980. A closed form equation for predicting the hydraulic conductivity of unsaturated soils. Soil Sci. Soc. Amer. J. 44: 892-898.

Van Grinsven, J.J.M., N. van Breemen and J. Mulder, 1987. Impact of acid atmospheric deposition on woodland soils in the Netherlands. I. Calculation of hydrochemical budgets. Subm. for publ. Soil Sci. Soc. Amer. J.

Van Vliet, B. and R. Langohr, 1981. Evidence of disturbance by frost of pore ferriargillans in silty soils of Belgium and Northern France, In: Bullock P. and C.P. Murphy (ed.) Soil Micromorphology, vol. 2, Soil genesis. Proceedings of the International Working Meeting on Soil Micromorphology, pp. 511-518.

VDI, 1983. Säurehältige Niederschläge: Enstehung und wirkungen auf terrestrische Oekosysteme. Verein Deutscher Ingenieure. Komm. Reinhaltung der Luft, Düsseldorf, 277 pp.

Visser, P.H.B. de, 1986. Inventories of standing tree crops and estimates of three growth in the past decades at three biogeochemical monitoring sites in the Netherlands. Internal Rept., Dept. of Soil Science & Geology, Agricultural University, Wageningen, the Netherlands.
Westhoff, V. and A.J. den Held, 1969. Plantengemeenschappen in Nederland, Thieme, Zutphen.
Winkels, H.J., 1985. Kwantitatieve aspecten van de koolstof en stikstof huishouding van de strooisellaag in de gronden van het Hackfortproject. Unpublished M.Sc. thesis, Dept. of Soil Science and Geology, Agricultural University, Wageningen, the Netherlands, 102 pp.
Witkamp, M., 1960. Seasonal fluctations of the fungus flora in mull and mor of an oak forest. Vd Wiel & co., Arnhem.

Appendix 1. Soil profile descriptions

Profile description Hackfort pit AI (south wall)

Described by: N. van Breemen, A.G. Jongmans and S. Slager, June 1980

AO, 3 – 0 cm; mainly litter of oak; moist; smooth and abrupt on:

A1, 0 – 8/11 cm; sandy loam, M50:210 – 300 μm; upper 3 cm 10YR2/2, with increasing depth changing into 10YR3/3; moist; less than 1% iron oxide concretions, very firm, diameter 1 – 3 mm, rounded; no macrostructure; fine biopores: 5 – 10/cm^2; fine and medium roots: 5 – 10/dm^2; friable; less than 1% gravel (diam. < 5 cm); clear and wavy on:

A3/B1, 8/11 – 21/30 cm; sandy loam, M50:210 – 300 μm; skeleton grains < 600 μm ; irregular tongues and spots of 10YR5/3 in the uppermost part; lower: 10YR5/8; moist; less than 1% iron oxide concretions, very firm, diameter 1 – 3 mm, rounded; no macrostructure; fine biopores: 10 – 15/cm^2; fine and medium roots: 5 – 10/dm^2; friable; less than 1% gravel (diam. < 3 cm); along roots locally A1 material; dead roots present; clear and wavy on:

B2, 21/30 – 44/50 cm; sandy loam, M50:210 – 300 μm, skeleton grains < 600 μm ; 7.5YR5/8, locally 7.5YR5/6 and 7.5YR7/8; moist; less than 5% iron oxide concretions, very firm, diam. < 1 cm, rounded, in a clustered distribution pattern; no macrostructure; fine biopores: 5 – 10/cm^2, locally 10 – 15/cm^2; fine and medium roots: 5 – 10/cm^2; friable to slightly firm (fragipan?); less than 1% gravel (diam. < 5 mm); clear and wavy on:

Clg, 44/50 – 74/78 cm; loamy sand, M50:150 – 210 μm; skeleton grains < 600 μm; moist; no single matrix colour; many (> 50%) coarse, diffuse, distinct, iron oxide mottles, 5YR5/8 to 10YR6/8, decreasing in abundance with depth; locally neoferrans (5YR5/8); reduction mottles, 10YR7/1 to 10YR6/1, 20 – 50%; disturbed stratification; fine biopores: 2 – 5/cm^2; few, fine, large and medium roots in a clustered distribution pattern; very friable (fragipan?); abrupt and wavy on:

C2g(G) > 74/78 cm; sand in layers of different texture, M50:150 – 210 μm and 210 – 300 μm; moist, changing to wet with increasing depth; 2.5Y5.5/0; few, coarse, diffuse, distinct to faint iron oxide mottles, 2.5Y6/4, abundance decreases with increasing depth; intact sedimentary stratification, parallel and locally inclined to the soil surface, boundary between these two stratifications is very sharp; fine and medium roots in a clustered distribution pattern: 0 – 1/dm^2; loose; strongly; calcareous

Remark: All horizons are present at all profile walls, in similar sequence. Locally, the A1-horizon is deeper and more pronounced. The profile has been described up to a depth of 120 cm.

Profile description Hackfort pit AII (south-west wall)

Described by: N. van Breemen and S. Slager on 24.06.1980

A0, 5 – 0 cm; oak leaves (5 – 3 cm); strongly decomposed oak litter from 3 – 0 cm; many fine (< 1 mm) roots; abrupt and wavy on:

A1, 0 – 5/10 cm; sandy loam, M50:105 – 210 μm; < 5% gravel, diam. < 3mm; 10 YR4/2, with few bleached grains in upper cm of horizon; where the profile is locally more sandy, the color of the upper 1 – 2 cm of the A1 is 10YR2.5/2, abruptly on 10YR4/2; no macrostructure, compact; less roots than in A0 and B21 (diam. < 3mm); friable to slightly firm; abrupt and wavy on:

B21, 5/10 – 15/20 cm; fine sandy loam, M50:210 – 300 μm; no gravel; 10YR5/8; no macrostructure; common fine and few large roots, many very large dead roots; friable; gradual and wavy on:

B22, 15/20 – 30/42 cm; sandy loam (but more clayey than B21); few quartz pebbles; 7.5YR5/7; no mottling; on the average ± 10% Fe-Mn concretions, 5 – 20 mm, in places clusters; weak, fine subangular blocky structure; common fine to large roots (diam. < 4 mm) in the upper 10 cm of the horizon, mainly growing horizontally; few fine to large roots in the lower part; slightly firm; locally (north-west wall) more sandy with very few Fe-Mn concretions; 10YR5/8; no macrostructure; common fine to large roots (diam. < 5 mm) throughout; clear and wavy on:

C11 g, 30/42 – 62/65 cm; fine sand; no gravel; 70% 2.5Y5.5/2 with few distinct, sharp, vertically elongated mottles, 5YR5/7; 25% very coarse, faint, diffuse, 10YR5/6 rust; no concretions; no macrostructure; biopores present; very few, mainly vertical roots, diam. < 3 mm; soft; no free carbonates; gradual and smooth on:

C12g, 62/65 – 85 cm; fine sand; no gravel; 2.5Y4/4 and 2.5Y5/1, approximately in 5 – 10 cm thick horizontal bands; no concretions; disturbed sedimentary stratification; biopores present; slightly fewer roots than C11g; soft; no free carbonates; very abrupt and wavy on:

C13g >85 cm; non-loamy fine sand; no gravel; 2.5Y5/1, locally slightly darker due to humic material in oblong vertical zones; no mottling; sedimentary stratification; no biopores; very few roots; soft; strongly calcareous; the material remains about the same to a depth of 180 cm.

Profile description Hackfort pit BI

Described by: A.G. Jongmans and S. Slager, May 1980

A0, 4 – 0 cm, (thickness varying between 3 and 8 cm over 2.5 m), mainly oak leaf litter, upper 3 cm dry, fresh and partly decomposed, lower 1 cm strongly decomposed material; herbs rooting in lower 1 cm; very sharp and slightly wavy on:

A1, 0 – 3 cm, (thickness varying between 2 and 6 cm over 2.5 m); fine sand to loamy fine sand (M50:210 – 300 μm); 10YR2/2, some bleached sand grains; moist; fine biopores present, but not counted; no macro-structure; few roots (diam. < 2 mm), distribution somewhat clustered; friable; abrupt (< 1 cm) and wavy on:

B21, 3 – 15 cm, (thickness varying between 8 and 19 cm over 2.5 m); fine sand to loamy fine sand (M50:210 – 300 μm); less than 5% gravel (diam. < 4 mm); moist; 10YR4/3, sometimes 5YR3/2 in the upper cm; no macrostructure; fine biopores present; loose fabric; roots present, mainly up to 2 mm in diam.; < 5% Fe-concretions, diam. < 4 mm; friable; clear and wavy on:

B22, 15 – 46 cm, (thickness varying between 27 and 40 cm over 2.5 m); fine sand (M50:210 – 300 μm); < 5% gravel, diam. < 1 mm; 10YR4/3 with 20 – 50% sharply bounded, rounded spots, diam. < 2 – 3 cm, of 10YR6/8; moist; less than < 5% Fe-concretions, up to 2 mm in diam.; no macro-structure; fine biopores present; many fine and large roots randomly distributed; friable; loose; lower few centrimetres of horizon contain loamy spots, presumably B2t-material; clear (2 cm) and wavy on:

B3t, 46 – 66 cm, (thickness varying between 18 and 22 cm over 2.5 m); fine sand (M50:210 – 300 μm), with discontinuous, strongly loamy lamella between 58 and 66 cm, loamy spots, presumably remnants of a lamella, between 49 and 52 cm, and at 54 cm a stone line with grains up to 5 mm; moist; 10YR8/8 in light-textured material, 7.5YR4/4 in lamellae; few, fine, faint iron mottles; disturbed sedimentary stratification; fine biopores present where lamellae are disturbed; few roots, in a clustered distribution pattern; soft where light-textured, slightly brittle where fine textured; less than 5% gravel (diam. < 3 cm); abrupt and wavy on:

C11g, 66 – 103 cm; alternation of non-loamy medium sand (M50:300 – 420 μm) and coarse sand (M50:420 – 600 μm); less than 5% gravel (diam. < 3 cm); moist to wet; 10YR7/2; 5 – 20%, faint, coarse, mottles, 10YR7/8, mainly at the upper side of the horizon; slightly disturbed sedimentary stratification; very few, fine biopores in a clustered distribution pattern; very few roots, clustered; soft; abrupt and smooth to slightly wavy on:

C12g, 103 – 130(+)cm; random alternation of medium sand (M50:420 – 600 μm) and sharply bounded, rounded and angular bodies (diam. < 20 – 30 cm), sometimes contorted lamellae of silty loam; wet; 10YR7/2 (sand) and N4/0 (silty loam); less than 5% gravel (diam. < 5 mm), associated with medium sand; coarse, diffuse, iron mottles, 10YR7/6; loamy bodies surrounded by 1 cm wide iron band, 5YR5/8; few to no roots (present roots concentrated in loamy material); soft (sand), non sticky, non plastic (silty loam); no free cabonates.

Profile description Hackfort pit BII

Described by: A.G. Jongmans, N. van Breemen and S. Slager, May, 1980

A0, 4 – 0 cm, (thickness varying from 2 to 6 cm over 2.5 m); mainly oak leaf litter; upper 3 cm fresh and partly decomposed, lower 1 cm strongly decomposed material; upper 3 cm dry, lower 1 cm moist; tree roots branch in to this horizon; abrupt and flat on:

A1(2), 0 – 4 cm, (thickness varying from 2 to 6 cm over 2.5 m); fine sand (M50:210 – 300 μm); less than 5% gravel, diam. < 3 mm; moist; 10YR2/2 with many bleached grains; few, fine, iron concretions (diam. < 3 mm); no macrostructure; fine biopores present; few fine roots, no larger roots; friable; abrupt and slightly wavy (sometimes tonguing) on:

B21(h), 4 – 8 cm, (thickness varying from 3 to 6 cm); fine sand (M50:210 – 300 μm); less than 5% gravel (diam. < 3 mm); moist; 5YR3/3 changing with depth to 5YR4/3; few, fine iron concretions (diam. < 3 mm); no macro-structure; fine biopores present; few fine and large roots; friable; in some places root channels filled-in with A1 (2) material; clear (2 – 3 cm) and slightly wavy on:

B22, 8 – 29 cm, (thickness varying from 13 to 21 cm over 2.5 m); fine sand (M50:210 – 300 μm); less than 5% gravel (diam. < 3 mm); 10YR5/6; few, fine iron concretions, diam. < 10 mm; 5 – 20%, irregular, faint, coarse mottles, 10YR6/6; moist; loose; fine biopores present; many fine and large roots (diam. < 10 mm), mainly horizontally; very friable; in some places filled-in root channels; abrupt and slightly wavy on:

B23, 29 – 48 cm, fine sand (M50:210 – 300 μm) with some strongly loamy spots (B2t-remnants?); less than 5% gravel, diam. < 3 mm; 10 YR7/8 with common, medium to coarse, distinct, irregularly shaped spots of 7.5YR5/6 (loamy material); few, fine to medium, faint iron oxide concretions; moist; no macrostructure; fine biopores present; many fine and large roots (diam. < 3 mm), mainly running in horizontal direction; (this horizon contains the largest number of roots); very friable to slightly brittle in loamy spots; in some places filled-in root channels; abrupt and slightly wavy on:

B31t, 48 – 60 cm; medium sand (M50:300 – 420 μm), discontinuous stoneline with grains up to 4 mm; discontinuous loamy lamellae (illuviated clay); moist; sand: 10YR7/6, changing with depth to 10YR8/4, loamy lamellae: 7.5YR5/6; disturbed sedimentary stratification; very few, fine biopores; very few roots, clustered distribution, mainly vertical; friable in light textured material, slightly firm in loamy material; abrupt and smooth on:

B32tg, 60 – 102/108 cm; light textured material with loamy lamellae (illuviated clay) of variable thickness which are absent locally; light-textured material: medium sand (M50:400 – 600 μm); lamellae: sandy loam; less than 5% gravel, diam. < 5 mm, in some places 5 – 20% gravel; moist; 10YR7/2 (light textured material) and 7.5YR5/8 (loamy lamellae); few, faint, irregular Fe-mottles, 10YR6/8; nearly undisturbed, sedimentary stratification; little or no fine biopores; few roots, clustered distribution; loose in light textured material, slightly brittle to soft in loamy lamellae; abrupt and slightly wavy on:

C1g, 102/108 – 120 cm; very fine sandy loam (M50:50 – 75 μm), alternating with horizontal layers of fine sand (M50:100 – 150 μm), no gravel; wet; 5Y5/2; common fine and coarse iron mottles in bands and around root channels: 7.5YR6/8 – 10YR6/8; nearly undisturbed, sedimentary stratification; in some places very few fine biopores; fine and medium roots (diam. < 3 mm), concentrated in upper 10 cm of this horizon, mainly running horizontally; soft; no free carbonates; abrupt and smooth on:

C2g, > 120 cm; calcareous, no roots.

By auger:
120 – 150 cm: calcareous, gray fine sand
150 – 160 cm: calcareous, dark gray silty loam
160 – 210 cm: calcareous, gray fine sand

Profile description Hackfort pit CI

Described by: N. van Breemen and S. Slager, June 1980

The pit CI contained two rather contrasting sections, i.e.
1. an undisturbed profile on the northern and western side of the pit;
2. a reworked profile on the eastern side of the pit.
The two profiles occurred side by side at the southern side.
In the upper part of both the undisturbed and reworked profile a micropodzol is present. The description refers to the undisturbed profile.

6/4 – 0 cm; oak leaf litter, upper 2 cm dry and undecomposed, lower 2 – 4 cm moist partly decomposed; the lower part of this horizon is rooted by herbs; abrupt and smooth on:

A1(2), 0 – 4/8 cm; non loamy fine sand with many bleached grains (M50:210 – 300 μm); moist; 10YR2/2; no macrostructure; fine biopores present; loose; few fine, no large roots; very friable; clear and wavy on:

B21(h), 4/8 – 8/15 cm; fine sand (M50:210-300 μm); moist; 10YR4/4; no macrostructure; fine biopores present; few roots, diam. up to 2 mm; friable; clear and wavy on:

B22, 8/15 – 28/35 cm; loamy fine sand (M50:210 – 300); moist; 7.5YR5.5/8; very few iron concretions, diam. < 5 mm; no macrostructure; fine biopores present; few fine and few large roots (diam. < 1 cm); many large dead roots (diam. < 2 cm); friable to slightly firm; gradual and wavy on:

B23(t), 28/35 – 60/65 cm; horizontal, partly disturbed, alternation of fine sand and loamy fine sand lamallae (M50:210 – 300 μm); moist; loamy material 7.5YR6/6, elsewhere 2.5Y6/4 – 10YR6/6; few iron concretions, diam. < 5 mm; disturbed sedimentary stratification; fine biopores present, only in the upper 10 cm; few fine and few large roots up to 1 cm; deeper no roots; friable; very few gravel, diam. < 1 cm; gradual and smooth on:

C1g, 60/65 – 200(+) cm; horizontal, slightly wavy, alternation of medium sand with gravel (diam. < 3 cm) and medium and coarse sand; in some places loamy bodies (diam. 2 – 20 cm); wet; 10YR3.5/7, changing with depth into 2.5Y6/2, in loamy bodies 5Y4/1; in the upper 15 cm many, coarse, faint, iron mottles; undisturbed sedimentary stratification; no biopores; no roots, except in loamy bodies; loose to non plastic, non sticky; free carbonates absent.

Profile description Hackfort pit CII

Described by A.G. Jongmans and N. van Breemen, June 1980
This soil appears to be reworked to a depth of 80 – 115 cm.

A0, 4 – 0 cm; (thickness varying from 3 to 6 cm); mainly oak leaf litter; decomposed lower part 5YR3/3; many fine (< 1 mm) roots; abrupt and wavy on:

A1(2), 0 – 8/13 cm; fine to medium sand (M50:210 – 420 μm); no gravel; 10YR2/1 with common bleached grains; no mottling; no macrostructure; moderate fine (< 1 mm), few medium (1 – 5 mm) roots; friable; abrupt and slightly wavy on:

B2(h), 8/13 – 10/25 cm; (thickness varying from 0 to 15 cm); medium sand (M50:300 – 420 μm); few gravel, diam. < 4 mm; 7.5YR4/3 (where the horizon is clear and thick) to 10YR5/6 (where the horizon is thinner); no mottling or concretions; no macrostructure; moderate fine and moderate medium roots; friable; clear and wavy on:

C11g, 10/25 – 32/65 cm; medium sand (M50:300 – 420 μm); 10YR7/4 to 10YR6/3; common, faint, coarse mottles, 10YR6/6, increasing in abundance with depth; no macro structure; common fine and many medium roots; very friable to loose; few gravel, diam. < 2 mm; occasional patches of loamy material; often slightly humic around roots; abrupt and wavy on C13g (see later); clear and wavy on C12g.

C12g; 33/40 – 48/60 cm; fine sand (M50:210 – 3000 μm); matrix 10YR7/4 to 10YR8/4, locally (< 5%) 10YR6/6 in bands of material with illuviated clay; common fine to coarse, faint mottles, 10YR5/8; no macrostructure; moderate fine, many medium roots; friable; no coarse fragments; sometimes slightly humic around roots; abrupt and irregular on:

C13g; 32/65 – 55/82 cm; medium sand (M50:300 – 420 μm); 10YR6/8; common 10YR2/1 Fe-Mn concretions, diam. 1 – 10 mm, hard and rounded; few gravel, diam. < 10 mm; few bands of 10YR6/6 (material with illuviated clay); no macrostructure; locally many roots and locally no roots; practically all thick roots (max. diam. 1 cm) seen in the pit are found in this horizon; friable; no free carbonates; abrupt and wavy or tongued on:

C14g, 48/83 – 80/115 cm; medium sand (300 – 420 μm); 2.5Y7/3 (20%) and 10YR7/6 (70%); common, large, distinct, diffuse mottles, 5YR5/6; moist; no macrostructure; locally fine to coarse roots, vertically in tongues; very friable; no coarse fragments; clear and wavy on:

C15g, > 80/115; fine, medium and coarse sand in disturbed sedementary layering;

5Y6/1.5, locally 5Y4/1 around dead roots below 120 cm depth; few, large, distinct, diffuse mottles, 10YR8/8 – 10YR7/8; wet; no macrostructure; few roots to at least 120 cm; few gravel, diam. < 5 mm; no free carbonates.

Profile description Hackfort pit D

Described by: N. van Breemen and S. Slager, June 1980

A1, 0 – 8/12 cm; fine sand (M50:210 – 300 μm); moist; 10YR2/1.5; granular structure; very loose with many fine biopores; abundant fine, medium and large roots, diam. < 1 cm; very friable; rich in free carbonates; clear and smooth on:

AC, 8/12 – 14/20 cm; fine sand (M50:210 – 300 μm); moist; 80% 10YR2.5/2 spotted with (sharply bounded, erratic) 20% 2.5Y6/2; no macrostructure, fine biopores present; common fine roots (< 3 mm), many roots between 3 and 8 mm; friable to slightly firm; rich in calcium carbonate; abrupt and smooth (in places tongued) on:

C1g, 14/20 – 50/60 cm; mixture of discrete, sharply bounded units (several dm across), with the following characteristics:

relative abundance	texture	M50:	$CaCO_3$	color, moist:
40%	fine sand	210-300 μm	+ +	2.5Y5/2
40%	medium sand	300-600 μm	–	10YR6/3
10%	fine sand	150-210 μm	–	2.5Y6/2
5%	fine sandy loam	–	+ +	5Y4/1
5%	fine sand	150-210 μm	+ +	10YR5/3

moist; no macro-structure; fine biopores present; roots (diam. 3 – 10 mm) mainly present in upper 10 cm of this horizon, loose to firm depending on texture; clear and broken on:

IIB21tb, 50/60 – 70/85 cm; more or less horizontal alternation of medium sand (M50:150 – 600 μm), 10YR7/2 and 30% loamy medium sand, 7.5YR5/8 (lamella of illuviated clay?); moist; 20 – 50%, coarse, faint, iron mottles (10YR6/6); no macrostructure; fine biopores present; no fine, few large roots (3 – 20 mm); loose and slightly sticky; less than 5% gravel, diam. < 5 mm; no free carbonates; abrupt and irregular on II B22sb or II C2gb.

IIB22sb, 70/85 – 75/110 cm; moist to wet sandy loam to loam; 7.5YR5/8; 20 – 30% fine, hard, Fe concretions; no macrostructure; fewer roots than in IIB21 (diam. 4 – 20 mm), sticky and plastic; no free carbonate; abrupt and broken on:

IIC2gb, 70/110 – 130 (+) cm; medium sand with discrete units of loam and of fine sand; wet; side by side 5Y4/1 and 2.5Y5/2; no macrostructure; in places roots up to 20 mm; soft; strongly calcareous.

Appendix 2. Soil micromorphological descriptions

The micromorphological description, made by Van Dis (1984), have been summarized in tabular form, indicating the abundance of the various features by bars:

none:	not present
broken:	present in small quantities
thin, continuous:	common
thick:	very common

The terminology used is according to Brewer (1964).

Appendix 2A. Micromorphology of soil profile A-1.

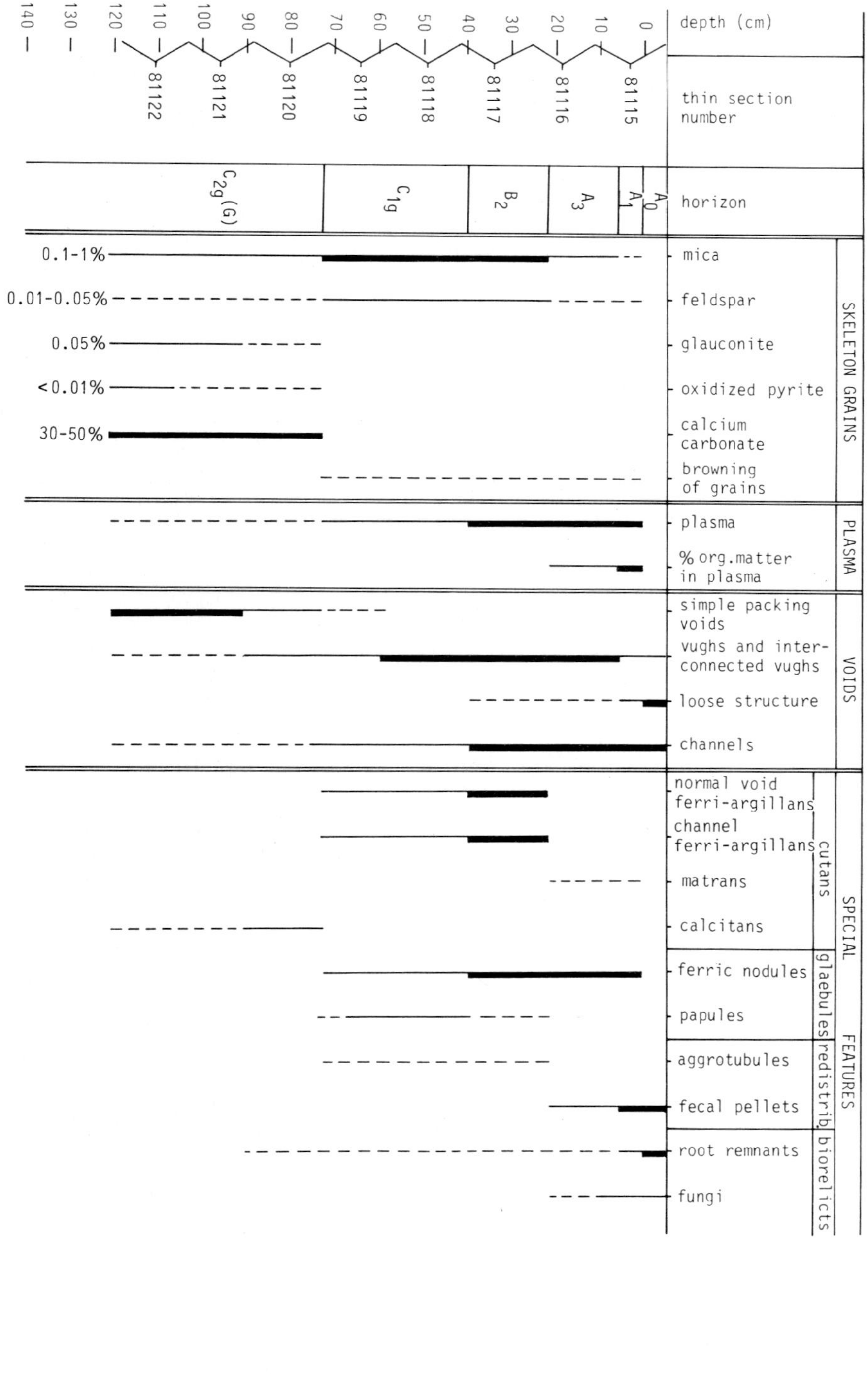

Appendix 2B. Micromorphology of soil profile B-1.

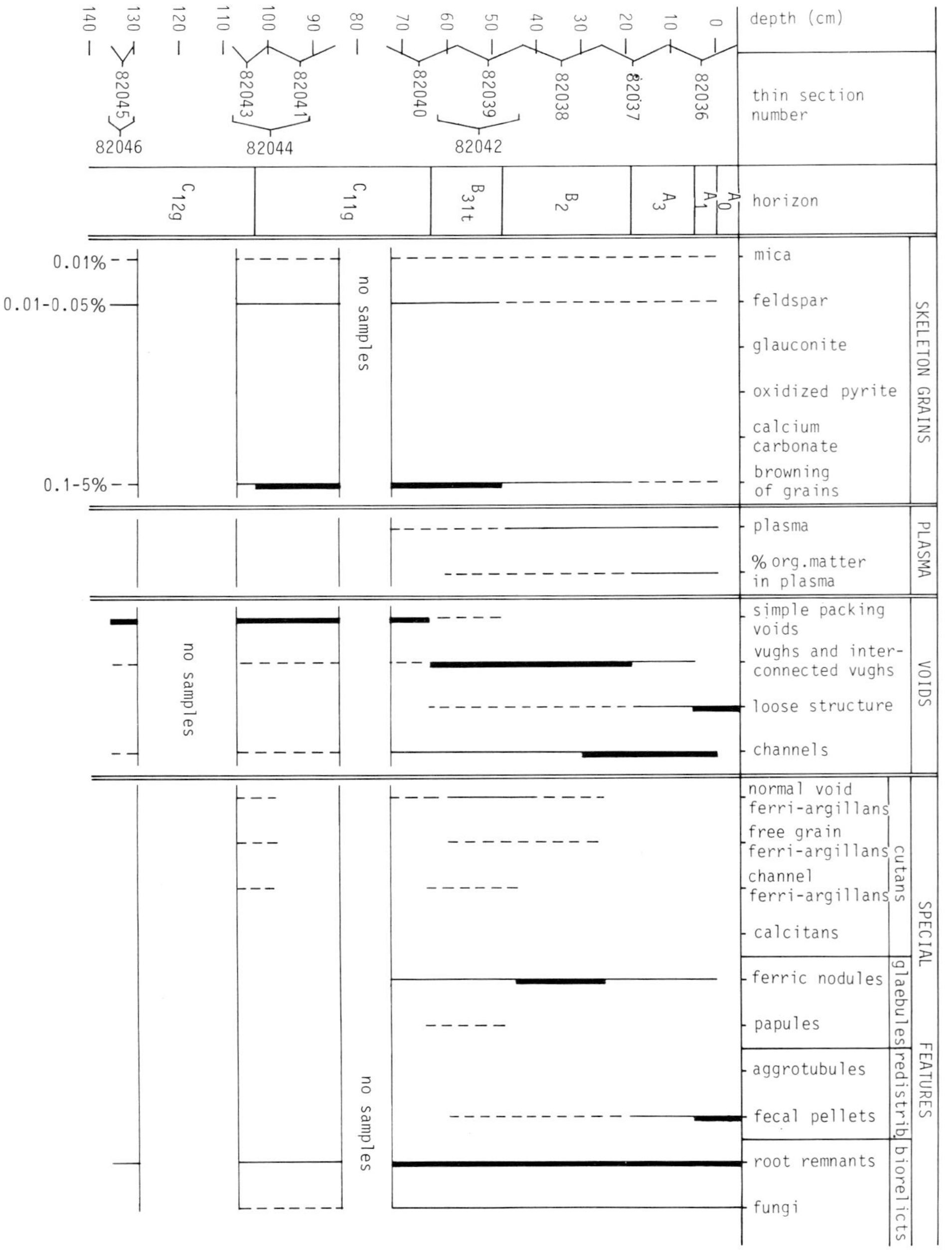

Appendix 2C. Micromorphology of soil profile C-2.

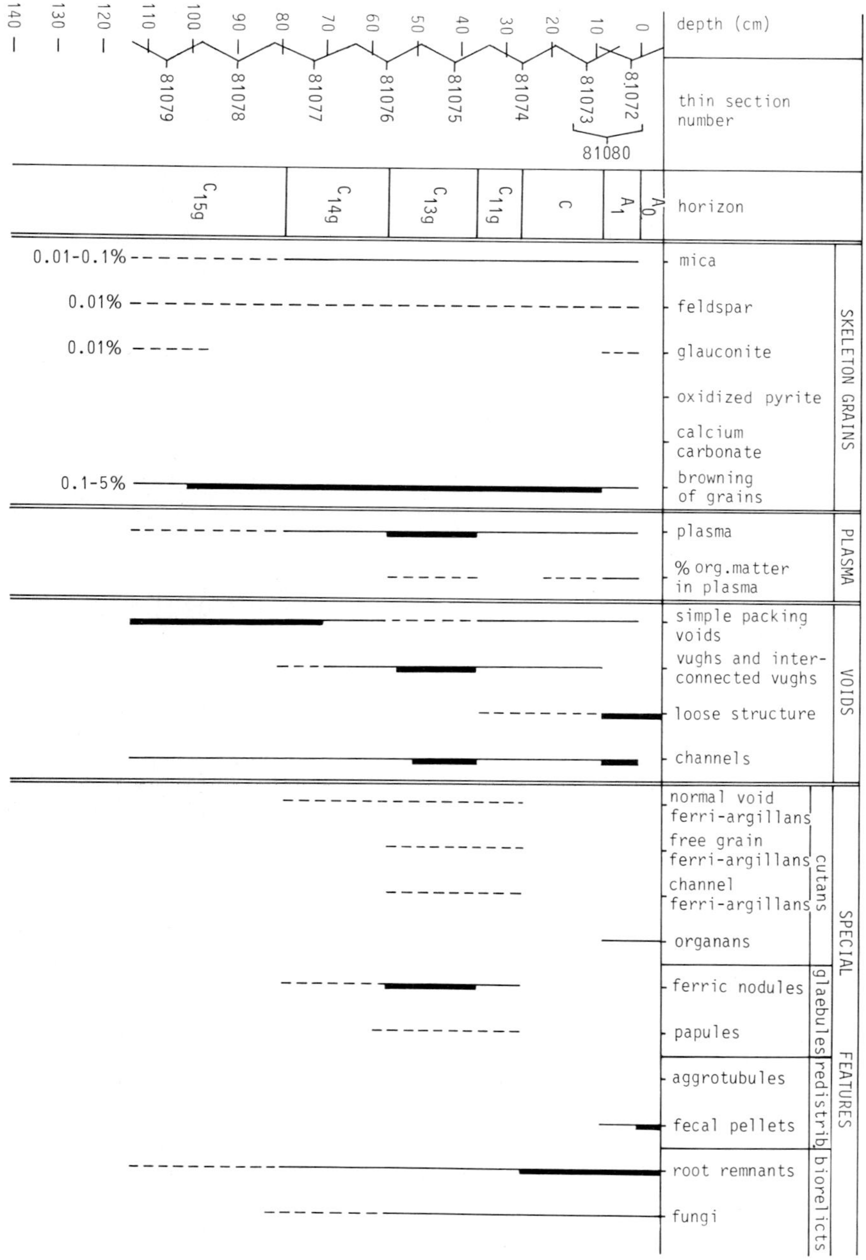

Appendix 2D. Micromorphology of soil profile D-1.

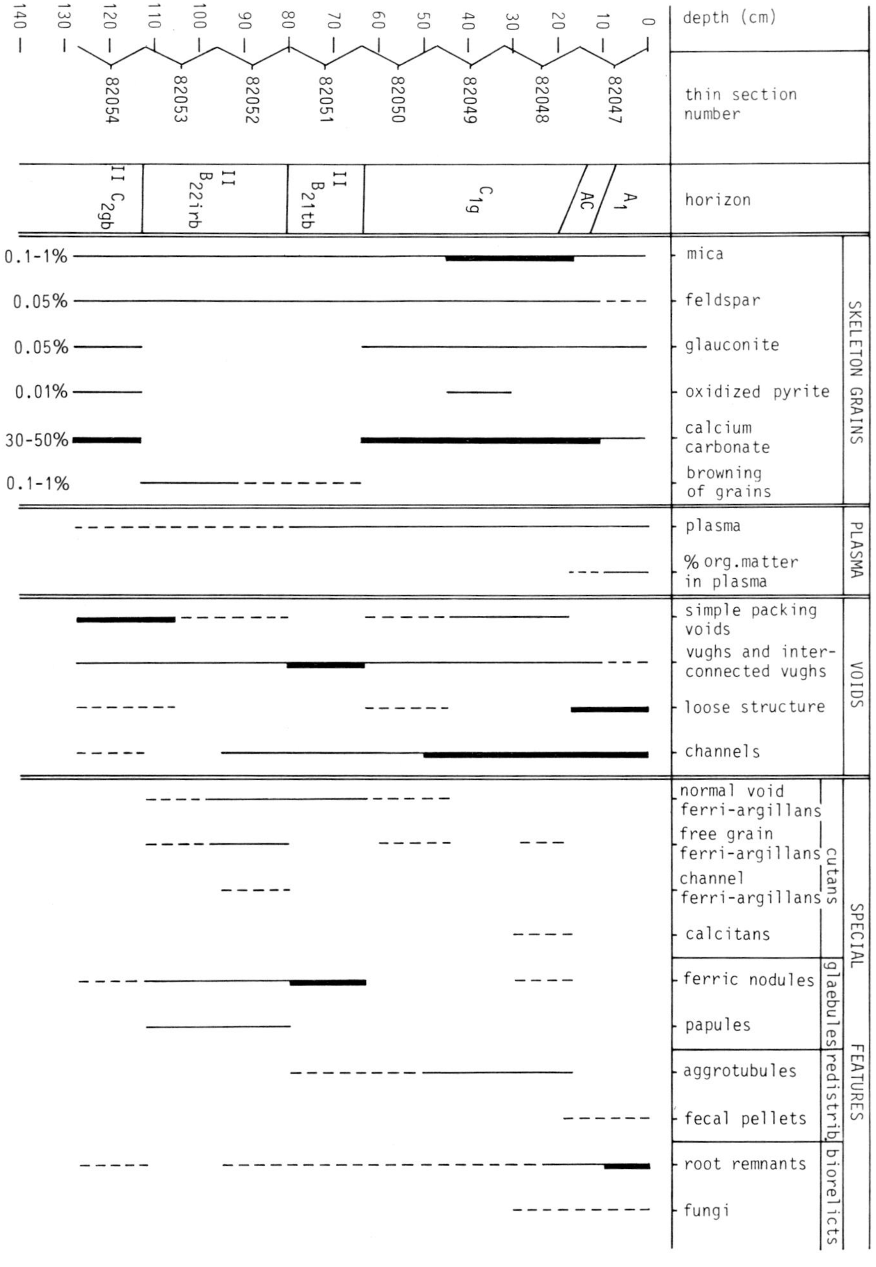

Appendix 3. Water retention data of soil samples

Water retention data of soil samples from each plot. Values are means and standard deviations of the volume fraction (%) of water present at the pressure potentials (mbar) indicated, as determined by desorption of water-saturated soil. The measurements were carried out by B. Kroesbergen of the Soil Tillage Laboratory, Agricultural University, Wageningen.

	Depth	number of samples	Pressure potential (mbar)							
			0	-10	-32	-100	-200	-500	-2500	-16000
A	0-5	12	59.7 7.4	50.1 2.8	44.5 3.3	36.3 3.4	30.6 2.4	26.6 1.7	20.5 0.4	17.6 0.7
	5-15	18	55.3 3.5	41.3 2.2	32.9 3.3	23.5 1.6	20.1 1.3	17.9 1.3	12.5 1.1	9.1 1.1
	15-25	18	49.5 3.4	40.1 2.3	31.9 1.1	24.0 1.9	20.9 2.0	18.8 2.0	13.8 1.6	10.7 1.5
	25-35	12	51.6 4.6	39.5 2.2	31.5 2.2	24.2 2.9	21.3 2.9	19.4 2.9	16.2 0.7	12.9 1.9
	35-45	12	46.2 2.4	38.1 1.6	33.0 1.5	24.8 1.9	21.0 1.9	18.9 1.8	15.5 1.1	12.4 1.1
	45-55	6	42.6 2.2	34.9 1.6	30.7 1.4	19.1 2.1	15.6 1.9	13.2 1.8	10.0	6.6
	55-65	12	43.6 1.6	38.0 1.9	34.6 3.3	18.9 4.9	14.3 3.7	11.7 3.1	6.8 0.6	4.5 1.1
	65-75	6	42.9 1.9	36.0 1.6	31.8 0.7	16.8 1.2	13.4 1.2	11.4 1.2	6.3	4.0
	75-85	6	40.9 1.6	39.3 1.1	37.8 1.2	18.9 3.3	12.5 2.0	9.1 1.2	4.2	3.0
	95-105	6	42.8 0.7	39.9 1.0	37.6 1.7	17.3 2.6	12.0 1.8	8.8 1.4	4.8	3.0
	110-120	6	39.0 0.9	38.9 0.6	38.2 0.7	9.6 0.9	5.7 0.6	4.4 0.5	4.0	2.4
B	0-5	12	67.7 7.5	56.3 7.3	49.2 7.6	38.3 8.8	30.9 7.1	26.8 6.6	24.8 3.7	19.5 2.2
	5-15	9	62.1 3.7	46.2 2.7	39.3 2.1	28.4 2.0	21.5 1.2	18.0 1.3	13.6 0.4	11.0 0.2
	15-25	12	58.8 3.8	44.9 2.5	37.5 2.8	26.1 2.3	20.7 1.3	17.3 1.1	11.2 0.9	8.8 0.2
	25-35	18	57.8 1.9	44.4 2.6	35.8 3.0	23.7 2.8	19.2 2.1	16.2 1.8	10.5 0.9	7.7 0.5
	35-45	12	53.9 3.3	41.4 2.4	34.0 2.8	22.4 3.1	17.7 2.7	14.8 2.8	10.4 0.1	7.4 0.1
	45-55	12	42.2 3.9	34.5 2.1	29.4 1.7	16.5 3.3	12.6 3.2	11.0 3.0	7.6 1.6	5.8 1.3
	55-65	8	40.9 3.1	33.1 2.4	27.8 1.6	14.7 2.8	11.3 2.5	9.7 2.5	6.9 3.2	5.1 3.3
	65-75	9	40.5 2.0	33.7 1.0	26.5 2.2	9.4 1.8	6.6 1.3	5.2 1.0	3.8 1.2	2.4 0.5
	75-85	6	40.0 2.0	33.3 1.1	24.0 1.8	7.9 0.7	5.6 0.4	4.3 0.4	3.8	2.3
	95-105	9	42.4 2.3	34.5 2.1	25.6 1.5	7.4 1.2	4.9 1.2	3.6 1.0	2.6 0.1	1.8 0.1
	110-120*	6	38.8 2.6	36.0 3.4	34.9 4.1	31.9 7.2	27.4 6.8	20.0 3.4	8.5 2.1	5.7 0.1
	110-120°	9	38.1 2.4	33.9 4.2	29.8 8.4	23.4 13.7	20.0 12.4	14.7 8.5	6.2 4.3	4.2 2.7
C	0-5	15	61.0 4.8	48.5 4.8	36.1 4.7	26.3 4.9	21.6 4.2	17.9 3.7	16.1 2.8	13.9 1.7
	5-15	9	53.4 2.7	43.2 1.8	32.5 2.1	17.4 2.4	13.8 2.0	11.3 1.6	8.7 1.9	6.7 1.5
	15-25	6	44.6 1.3	29.4 1.4	16.2 1.9	6.8 0.5	4.9 0.5	3.7 0.5	3.0	2.1
	25-35	11	45.1 1.5	37.7 1.8	30.3 2.7	12.2 1.7	8.3 1.4	6.8 1.4	5.7 0.6	3.8 0.6
	35-45	12	44.9 2.5	38.8 3.1	25.7 1.8	14.8 4.1	12.6 4.0	10.8 3.4	6.8 3.6	5.0 3.5
	45-55	18	43.1 1.2	36.0 0.8	28.5 2.9	13.1 2.6	9.4 2.1	7.9 1.9	4.4 1.4	3.2 0.2
	65-75	18	41.3 1.8	35.1 1.7	28.0 4.6	16.5 7.7	13.1 7.6	11.3 7.3	7.3 5.0	5.5 4.2
	110-120	6	35.3 2.0	33.7 2.1	33.1 2.2	20.8 3.7	11.7 3.0	8.0 2.1	4.6	2.9
D	0-5	3	59.8 1.0	52.5 0.3	40.9 0.8	31.6 1.1	26.6 1.1	22.4 0.7		
	11-16	3	54.9 3.7	43.6 2.7	32.3 2.6	18.4 3.2	14.8 2.5	11.8 1.9		
	25-30	3	46.2 1.6	39.4 2.4	28.4 2.3	12.7 1.2	9.6 0.8	8.3 0.8		
	45-50	6	43.2 1.2	36.0 0.7	30.1 0.6	14.2 0.8	9.3 0.6	7.7 0.7		
	70-75	4	41.4 1.9	34.9 1.8	30.7 2.5	20.6 5.9	16.6 6.9	14.6 6.8		
	110-115	6	35.3 2.0	33.7 2.1	33.1 2.2	20.8 3.7	11.7 3.0	8.0 2.1		

* excluding coarse sandy samples

° including coarse sandy samples

– no volume-weighted sampling

Appendix 4. Fungi

Appendix 4A	Fungi of 3 plots in Hackfort in 1982-1984
Appendix 4B	Fungi observed by Witkamp in 1957-1958
Appendix 4C	Comparison of fungi reported in 1957-1958 and 1982-1984
Appendix 4D	Fungi collected in 1981-1984 and conserved in the herbarium of the Biological Station at Wijster

Appendix 4A. Fungi of 3 plots in Hackfort ('Oude Maat') in 1982-1984.

The presence of a species is given per plot, per relevé. Numbers are counted numbers of carpophores (fruitbodies). Symbols are estimates of the number of carpophores:

r = rare, 1 or 2 carpophores
oc = occasional, 3 to 9 carpophores
f = frequent, 10 to 29 carpophores
vf = very frequent, 30 to 100 carpophores
a = abundant, 100 to 500 carpophores
va = very abundant, more than 500 carpophores
x = present, but the number of carpophores is not counted or estimated.

In species with very small fruitbodies of only one or a few mm, not the number of fruitbodies is counted or estimated but the number of localities.
Species occurring outside but very close to a plot, are mentioned in brackets, e.g. [x] or [2], which means occurring, resp. occurring with 2 carpophores just beside the plot. Data between () refer to groups of fruitbodies, given if it is not possible to distinguish individual fruitbodies.
Identifications that are not certain, are indicated with 'cf'.

Plot 1					
year	82	82	83	83	84
month	10	11	10	11	10
day	14	10	13	1	24
relevé nr.	1	4	7	10	13
Agrocybe firma	1			2	
Amanita muscaria				[1]	
Antrodia semisupina		1			
Armillariella ostoyae		5	380		15
Ascocoryne cylichnium	x	(1)	50	(1)	
Bjerkandera adusta					(1)
Cerocorticium confluens					x
Chondrostereum purpureum					x
Clavariadelphus junceus				5	1
Clitocybe diatreta	4				
C. gibba	1		[x]		
C. metachroa			[3]		
C. phyllophila			[11]	[4]	[12]
C. vibecina					[x]
Collybia butyracea	[x]	[x]	[1]	[3]	[1]
C. cookei					85
C. dryophila	oc		[1]	1	
C. peronata	2		[x]		[x]
Conocybe macrocephala	1	1	5		10
C. rickenii				4	
Coprinus disseminatus	f				17
C. lagopus	3		8	1	2
Cortinarius alnetorum		33			
C. delibutus	1				
C. sertipes	13		6	15	17
Cylindrobasidium evolvens					x
Entoloma molliusculum		1			
E. turbidum	x				
Flagelloscypha minutissima					x
Galerina ampullaceocystis	1				
G. heimansii			1		
Ganoderma applanatum	1				
Hymenoscyphus immutabilis	x		x	a	x
H. rubicola				x	
H. scutula	x		(3)	x	x
Hyphoderma praetermissum					x
H. pallidum					x
Hypholoma fasciculare	r	21	25	12	[40]
Incrustoporia nivea			(1)		(1)
Inocybe cf cookei					1
Kneifiella sambuci					x
Kuehneromyces mutabilis		12	30		
Lactarius obscuratus		1			
L. quietus	r	[x]	[x]	[x]	
L. theiogalus	f	2	[2]	[1]	[4]
Lepista inversa					[35]

relevé nr.	1	4	7	10	13
L. nebularis		2	[12]	3	[13]
L. nuda	1	2		[5]	
Marasmiellus ramealis	a	4	40		42
Marasmius torquescens	2		3		
Melanoleuca melaleuca	r		5	2	
Merulius tremellosus		(1)			[x]
Mollisia spec.					x
Mycena filopes			2		3
M. flavoalba					1
M. galericulata	vf	17	85	36	53
M. galopoda	a	1	26	12	40
M. leptocephala			2		
M. mucor		1			3
M. pearsoniana					1
M. polyadelpha		5		43	
M. polygramma		3	7	2	5
M. pura	[x]		[4]		2
M. sanguinolenta	f		1		
M. smithiana		4			
M. speirea	5	2	22	5	9
M. stylobates			1		1
M. vitilis	f	9	39	15	30
M. vitrea	oc	1	1		7
Mycenella margaritispora					2
Naucoria escharoides	70	6	2	40	40
Nectria spec.	x	(6)		(4)	x
Oudemansiella radicata					1
Paxillus involutus	3			1	2
Peniophora spec.					x
Perichaena corticalis					x
Peziza cf badia	x				
Pezizella alniella		x			
Phallus impudicus	[x]	[x]	[x]		
Phlebia radiata			(1)	(2)	
Piptoporus betulinus		3	28	28	40
Pleurotellus herbarum					11
Pluteus atricapillus			1		
P. thomsonii			2		
Rickenella fibula	1		1	4	
Ripartites tricholoma					1
Russula cf nitida		1			
cf Rutstroemia firma					1
Schizopora paradoxa				(1)	
S. phellinoides					x
Scleroderma citrinum			1		
Sistotrema brinkmannii	x				
Stereum hirsutum	x				
Stereum rugosum				x	
Trechispora mutabilis					x
Tubaria furfuracea	x		7		

relevé nr.	1	4	7	10	13
Typhula erythropus	r		3	20	6
T. phacorrhiza					x
T. sclerotioides		7	20		
Xerocomus cf badius	r				
Xylaria hypoxylon	x	(12)	(5)	(10)	(6)
X. longipes			(1)		
Zignoëlla ovoida					x
Number of species just outside the plot:	3	3	10	5	9
Number of species in each relevé:	43	33	47	35	58

Total 103 species

Plot 2	year	82	82	83	83	84
	month	10	11	10	11	10
	day	14	10	13	1	24
	relevé nr.	2	5	8	11	14
Amanita rubescens					1	
Armillariella ostoyae			102	156	1	51
Ascocoryne spec.			[x]			
Bjerkandera adusta					x	
Calocera cornea					10	
Ceriporia excelsa			[x]			
Cerocorticium confluens			x			
Clitocybe candicans				2		13
C. diatreta						3
C. ditopa		1		1		
C. gibba			1	7	1	1
C. metachroa		oc	53	6	41	123
C. vibecina			64		12	2
Clitopilus hobsonii		oc				f
Collybia butyracea		4	41	7	6	48
C. cookei				72	15	10
C. dryophila		oc		2		
C. peronata		f		9	2	5
Cortinarius sertipes		23	60	39	15	35
Crepidotus variabilis		oc	7	30	[x]	oc
Dacrymyces stillatus		x	x	(2)	x	(1)
Entoloma juncinum					1	
Galerina cinctula				1		1
G. hypnorum		2	1		1	
Hymenoscyphus scutula		x	x	x		
Hypholoma fasciculare		2	93	160	24	111
Laccaria proxima				[6]	1	
Lactarius necator		1	[x]			
L. quietus		oc	1	2	2	
L. theiogalus		a	34	32	40	7
Macrotyphula fistulosa						1
Marasmiellus ramealis		a	25	14	62	1
Marasmius epiphylloides				9	3	22
Mollisia sp.			x			
Mycena cinerella					2	83
M. galericulata		vf	41	280	74	72
M. galopoda		a	2	12	10	13
M. haematopoda		oc				
M. metata		1				
M. cf polyadelpha			3			
M. polygramma			1			1
M. sanguinolenta		a		4	3	4
M. vitilis		oc	5	19	7	24
Nectria spec.		oc	(11)	(8)	(8)	x
Oudemansiella platyphylla		oc		2	2	1
Paxillus involutus		f	3	2		
Peniophora quercina					x	

relevé nr.	2	5	8	11	14
Phallus impudicus	oc		1		
Phlebia radiata	oc	x			
Piptoporus betulinus	[vf]	[x]	[x]	[x]	
Pleurotellus herbarum				2	vf
Pluteus atricapillus	r			1	
Polyporus brumalis					2
Protocrea delicatula		x			
Psathyrella fulvescens	5	3	1		3
P. hydrophyla	20	20	19	24	
P. spadicea		6			
P. spadiceogrisea				3	
P. squamosa		2	1		1
Psilocybe crobula			1	2	
Resupinatus trichotis	20				
Russula emetica	oc	2			
R. ochroleuca		3	6	1	
Rutstroemia sydowiana	x				
Schizopora paradoxa	f	(2)	(2)	(4)	(3)
Sphaerobolus stellatus	oc	[x]		(1)	
Stereum hirsutum	x		x		
S. rugosum					(1)
Stropharia aeruginosa ss.	3		1	2	1
Trechispora farinacea		x			
T. vaga				x	
Typhula phacorrhiza					x
Xerocomus chrysenteron	1		3		
Xylaria hypoxylon		x	x	[x]	
Number of species just outside the plot:	1	5	2	3	
Number of species in each relevé:	39	39	39	41	35

Total 74 species

Plot 3	year	82	82	83	83	84
	month	10	11	10	11	10
	day	14	10	13	1	24
	relevé nr.	3	6	9	12	15
Achroomyces peniophorae		x				
Amanita rubescens		1				
Armillariella ostoyae			62	68	2	48
Ascocoryne sarcoides		x				
A. spec. (imperfect)						x
Bjerkandera adusta						x
Botryobasidium subcoronatum					x	x
Calocera cornea		r		x	20	6
Cerocorticium confluens		x				x
Clitocybe candicans		[1]	[7]	5		3
C. gibba				8	2	
C. metachroa		vf	31	8	98	130
C. vibecina		5	38			
Clitopilus hobsonii				10		
Collybia butyracea		[7]	1	15	4	16
C. cookei						4
C. dryophila		vf		4	1	
C. peronata		a	5	100	32	24
Dacrymyces stillatus		oc		x	x	x
Galerina hypnorum		10			4	
Gymnopilus penetrans				x		
Hyphoderma praetermissum		x			x	x
H. puberum					2	x
Hypholoma fasciculare				21	6	8
Inocybe napipes		r				
Lactarius necator		[x]				1
L. quietus		vf	1	17	8	1
L. theiogalus		oc		4		
Marasmiellus ramealis					16	x
Mollisia spec.			x			x
Mycena cinerella						7
M. galericulata		vf	43	520	96	130
M. galopoda		va	1	90	40	110
M. inclinata					30	
M. rorida		r				
M. sanguinolenta		oc		1	3	10
M. stylobates				5		
M. vitilis		f	1	25	18	60
Nectria spec.		f	x			
Oudemansiella platyphylla		10		8	2	4
Peniophora quercina				x		
P. violaceolivida					x	
Phallus impudicus		oc				
Phanerochaete sordida		x				
Phlebia radiata		x	x	x	f	f
Piptoporus betulinus				[x]		
Pluteus atricapillus				x	1	1

relevé nr.	3	6	9	12	15
Psathyrella fulvescens	2		1		2
P. hydrophila	f	8		40	1
Psilocybe crobula	2		2	4	
Resupinatus applicatus			x	3	
Russulae ochroleuca	oc		18	4	4
Rutstroemia firma	2				2
Schizopora paradoxa	x	x		vf	x
S. phellinoides		x			x
Scleroderma citrinum	f		5	2	x
Sphaerobolus stellatus					40(1)
Stereum hirsutum					40(2)
S. rugosum	r				17
Stropharia aeruginosa ss.	11		4	2	
Trametes versicolor	[x]				25(2)
T. spec. (juveniel)				4	
Trechispora cohaerens	x			x	x
T. farinacea				x	x
Xenasma filicinum					x
Xerocomus badius	6		4		
X. chrysenteron	1				
Xylaria hypoxylon	[x]	[x]			
Zignoëlla cf ovoidea					x
Number of species just outside the plot:	5	2	1		
Number of species in each relevé:	42	16	31	34	42

Total 69 species

Appendix 4B. Fungi observed by Witkamp in 1957 and 1958.

After Witkamp (1960) and Minderman (1981). Per humus type per year presence of a species is indicated with –, or the frequency (number of visits, out of 13 paid, during with the species was observed) is indicated.

Some identifications are questionable. These are shortly discussed below.

– *Russula rubra* and *Lactarius mitissimus* are species of coniferous woods (Moser, 1983). So their presence in this oak coppice is unlikely. Maybe *R. emetica*, or a closely related species, and *L. theiogalus* are meant instead.

– *Hygrophoropsis aurantiaca* (Dutch name: 'valse hanekam' or 'valse cantharel') is mentioned by Witkamp (1960) but not by Minderman (1981). *Hygrocybe cantharellus* (Dutch name: 'trechterwasplaat') is mentioned by Minderman but not by Witkamp. I included both names in this table, but these two names probably refer to the same species. However, their presence in this oak coppice is unlikely: *Hygrophoropsis aurantiaca* is a species of coniferous woods, *Hygrocybe cantharellus* a species of bogs and wet woods. Minderman mentioned also all the Dutch names; he stated *H. cantharellus* as 'hanekam'. This, however, is the Dutch name of *Cantharellus cibarius*. So it is possible that neither *H. aurantiaca* nor *H. cantharellus* were observed here, but *C. cibarius*. This is in agreement with the habitat preference of *C. cibarius*.

– *Calocybe ionides* is a very rare species of calcareous soils. This species was reported in all of the three plots. Being a calciphilous species, the presence in at least the two plots with acid soils is very unlikely. Confusion with *Lepista nuda*, a common species of acid woods, is possible. Therefore identification is probably wrong.

– *Collybia confluens* is a species growing on litter on rather nutrient-rich soils. In 1982/84 *C. confluens* was not found. It is possible that *C. peronata*, which was invariably present in 1982/84, was wrongly identified as *C. confluens* in 1957/58.

– *Lactarius blennius* is a mycorrhizal fungus growing with beech (*Fagus sylvatica*), which does not grow on the plots. Probably *L. necator* was mistakenly identified as *L. blennius*.

– Identifications of *Mycena corticola*, *M. epipterygia*, *Cliticybe brumalis*, *Neotiella rutilans*, *Lycoperdon perlatum*, *Laccaria laccata*, *Armillariella mellea*, *Aminata vaginata* and of all *Cortinarius* species should be checked. This is, however, not possible as no collections have been conserved. *Clitocybe brumalis* probably refers to *C. vibecina*, *L. perlatum* is probably *L. foetidum*, *Laccaria laccata* probably *L. proxima, Amanita vaginata* most likely *A. fulva* and *Armillariella mellea* probably *A. ostoyae*.

humus type:	calc. mull		acid mull		mor	
year:	57	58	57	58	57	58
species:						
1 Mycena galopoda	5	3	6	4	5	4
2 Scleroderma citrinum	6	1	5	2	6	6
3 Stropharia aeruginosa	7	5	3	2	4	3
4 Laccaria laccata	5	4	5	1	5	3
5 Laccaria amethystina	4	3	1	2	6	2
6 Collybia dryophila	2		3	2	5	3
7 Amanita phalloides	1	1	2	1	2	7
8 Amanita citrina	2	1	1	2	3	5
9 Collybia confluens	2	2	4	1	4	1
10 Paxillus involutus	1	5	3	2		1
11 Collybia butyracea	2	1	3	–	3	2
12 Clitocybe brumalis	1		4	3	1	2
13 Mycena sanguinolenta	1		4	–	4	
14 Calocybe ionides	1	1	3		1	1
15 Mycena epipterygia	1		2	1	2	
16 Oudemansiella radicata		1	2	1	2	
17 Lactarius chrysorrheus		1	3		2	
18 Cortinarius paleaceus		1	1	2	1	
19 Oudemansiella platyphylla		1	2		1	1
20 Coprinus disseminatus		–	1		1	
21 Mycena galericulata	–		–		–	
22 Armillariella cf. ostoyae	–		–		–	
23 Lycoperdon perlatum	3	1	1			
24 Macrotyphula fistulosa	2	1		2		
25 Amanita muscaria		1	1	1		
26 Crepidotus variabilis	–		–			
27 Cortinarius multiformis	–		–			
28 Tricholoma sulphureum	–		–			
29 Xylaria hypoxylon	–		–			
30 Inocybe geophylla	4	1			–	
31 Marasmius rotula	–				–	
32 Lactarius vellereus			2	2	4	3
33 Amanita vaginata			2	1	4	2
34 Xerocomus subtomentosus			2	2		2
35 Lactarius subdulcis			2		3	
36 Xerocomus badius			1		1	1
37 Russula nigricans			1		3	
38 Psathyrella hydrophila			1	1	1	
39 Russula delica				2		1
40 Hypholoma fasciculare			–		–	
41 Hypholoma sublateritium			–		–	
42 Clavariadelphus junceus	2	2				
43 Inocybe asterospora	1	2				

	humus type:	calc. mull		acid mull		mor	
	year:	57	58	57	58	57	58
species:							
44	Mycolachnea hemisphaerica	1	1				
45	Clavulina rugosa	1	1				
46	Leucoscypha rutilans		2				
47	Scleroderma verrucosum		2				
48	Peziza badia	2					
49	Helvella macropus	2					
50	Hebeloma crustuliniforme	–					
51	Polyporus brumalis	–					
52	Lactarius necator			4	2		
53	Lactarius blennius			2			
54	Lactarius vietus			2			
55	Cortinarius alboviolaceus			–			
56	Russula laurocerasi			–			
57	Mycena corticola			–			
58	Tricholomopsis rutilans			–			
59	Russula fragilis					4	4
60	Dermocybe cinnabarina					3	4
61	Lactarius quietus					1	5
62	Russula cyanoxantha					1	3
63	Cortinarius hemitrichus					1	2
64	Russula emetica					3	
65	Hygrophoropsis aurantiaca					1	–
66	Hygrocybe cantharellus					–	
67	Marasmius androsaceus					–	
68	Collybia maculata					–	
69	Asterophora lycoperdoides					–	
70	Russula rubra					–	
71	Cortinarius elatior					–	
72	Lactarius mitissimus					–	
73	Lactarius camphoratus					–	
74	Mycena polygramma					–	

Appendix 4C. Comparison of fungi reported in 1957-1958 and 1982-1984.

a. Fungal species reported again, not reported again and newly found in the total surveyed area.

reported again:

Amanita muscaria
Armillariella cf. ostoyae
Clavariadelphus junceus
Collybia butyracea
C. dryophila
Coprinus disseminatus
Crepidotus variabilis
Hypholoma fasciculara
Lactarius necator
L. quietus
Macrotyphula fistulosa
Mycena galericulata
M. galopoda
M. polygramma
M. sanguinolenta
Oudemansiella platyphylla
O. radicata
Paxillus involutus
Peziza badia
Polyporus brumalis
Psathyrella hydrophila
Russula emetica
Scleroderma citrinum
Stropharia aeruginosa
Xerocomus badius
Xylaria hypoxylon

not reported again:

Amanita citrina
A. phalloides
A. vaginata
Asterophora lycoperdoides
Calocybe ionides
Clavulina rugosa
Clitocybe brumalis
Collybia confluens
C. maculata
Cortinarius alboviolaceus
C. elatior
C. hemitrichus
C. multiformis
C. palaeceus
Dermocybe cinnabarina
Hebeloma crustuliniforme
Helvella macropus
Hygrocybe cantharellus
Hygrophoropsis aurantiaca
Hypholoma sublateritium
Inocybe asterospora
I. geophylla
Laccaria amethystina
L. laccata
Lactarius blennius
L. camphoratus
L. chrysorrheus
L. mitissimus
L. subdulcis
L. vellereus
L. vietus
Leucoscypha rutilans
Lycoperdon perlatum
Marasmius androsaceus
M. rotula
Mycena corticola
M. epipterygia
Mycolachnea hemisphaerica
Russula cyanoxantha
R. delica
R. fragilis
R. laurocerasi
R. nigricans
R. rubra
Scleroderma verrucosum
Tricholoma sulphureum
Trichomolopsis rutilans
Xerocomus subtomentosus

newly found:

Achroomyces peniophorae
Agrocybe firma
M. filopes
M. flavoalba

Amanita rubescens
Antrodia semisupina
Ascocoryne cylichnium
A. sarcoides
A. spec. (imperf.)
Bjerkandera adusta
Botryobasidium subcoronatum
Calocera cornea
Ceriporia excelsa
Cerocortitium confluens
Chondrostereum purpureum
Clitocybe candicans
C. diatreta
C. ditopa
C. gibba
C. metachroa
C. phyllophila
C. vibecina
Clitopilus hobsonii
Collybia cookei
C. peronata
Conocybe macrocephala
C. rickenii
Coprinus lagopus
Cortinarius alnetorum
C. delibutus
C. sertipes
Cylindrobasidium evolvens
Dacrymyces stillatus
Entoloma juncinum
E. molliusculum
E. turbidum
Flagelloscypha minutissima
Galerina ampullaceocystis
G. cinctula
G. heimansii
G. hypnorum
Ganoderma applanatum
Gymnopilus penetrans
Hymenoscyphus immutabilis
H. rubicola
H. scutula sl.
Hyphoderma praetermissum
H. puberum
H. pallidum
Incrustoporia nivea
Inocybe cf. cookei
I. napipes
Kneifiella sambuci
Kuehneromyces mutabilis
Laccaria proxima
Lactarius obscuratus
M. haematopoda
M. inclinata
M. leptocephala
M. metata
M. mucor
M. pearsoniana
M. polyadelpha
M. pura
M. rorida
M. smithiana
M. speirea
M. stylobates
M. vitilis
M. vitrea
Mycenella margaritispora
Naucoria eschariodes
Nectria spec. (imperf.)
Oudemansiella badia
Peniophora quercina
P. violaceolivida
Perichaena corticalis
Pezizella alniella
Phallus impudicus
Phanerochaete sordida
Phlebia radiata
Piptoporus betulinus
Pleurotellus herbarum
Pluteus atricapillus
P. thomsonii
Protocrea delicatula
Psathyrella fulvescens
P. spadicea
P. spadiceogrisea
P. squamosa
Psilocybe crobula
Resupinatus applicatus
R. trichotis
Rickenella fibula
Ripartites tricholoma
Russula nitida
R. ochroleuca
Rutstroemia firma
R. sydowiana
Schizopora paradoxa
S. phellinoides
Sistotrema brinkmannii
Sphaerobolus stellatus
Stereum hirsutum
S. rugosum
Trametes spec. (juv.)
T. versicolor
Trechispora cohaerens

L. theiogalus
Lepista inversa sl.
L. nebularis
L. nuda
Marasmiellus ramealis
Marasmius epiphylloides
M. torquescens
Melanoleuca melaleuca
Merulius tremellosus
Mollisia spec.
Mycena cinerella
T. farinacea
T. mutabilis
T. vaga
Tubaria furfuracea
Typhula erythropus
T. phacorrhiza
T. sclerotioides
Xenasma filicinum
Xerocomus chrysenteron
Xylaria longipes
Zignoëlla ovoidea

b. Fungal species reported again, not reported again and newly found in plot 1.

reported again:

Amanita muscaria
Armillariella cf. ostoyae
Clavariadelphus junceus
Collybia butyracea
C. dryophila
Corpinus disseminatus
Mycena galericulata
M. galopoda
M. sanguinolenta
Oudemansiella radicata
Paxillus involutus
Peziza badia
Scleroderma citrinum
Xylaria hypoxylon

not reported again:

Amanita citrina
A. phalloides
Calocybe ionides
Clavulina rugosa
Clitocybe brumalis
Collybia confluens
Cortinarius multiformis
C. palaeceus
Crepidotus variabilis
Hebeloma crustuliniforme
Helvella macropus
Inocybe asterospora
I. geophylla
Laccaria amethystina
L. laccata
Lactarius chrysorrheus
Leucoscypha rutilans
Lycoperdon perlatum
Macrotyphula fistulosa
Marasmius rotula
Mycena epipterygia
Mycolachnia hemisphaerica
Oudemansiella platyphylla
Polyporus brumalis
Scleroderma verrucosum
Stropharia aeruginosa
Tricholoma sulphureum

newly found:

Agrocybe firma
Bjerkandera adusta
Clitocybe diatreta
C. gibba
C. metachroa
C. phyllophila
C. vibecina
Collybia cookei
C. peronata
Conocybe macrocephala
Mycena vitrea
Mycenella margaritispora
Naucoria escharoides
Phallus impudicus
Phlebia radiata
Piptoporus betulinus
Pleurotellus herbarum
Pluteus atricapillus
P. thomsonii
Rickenella fibula

C. rickenii
Coprinus lagopus
Cortinarius alnetorum
C. delibutus
C. sertipes
Entoloma molliusculum
E. turbidum
Galerina ampullaceocystis
G. heimansii
Ganoderma applanatum
Hypholoma fasciculare
Inocybe cf. cookei
Kuehneromyces mutabilis
Lactarius obscuratus
L. quietus
L. theiogalus
Lepista inversa sl.
L. nebularis
L. nuda
Marasmiellus ramealis
Marasmius torquescens
Melanoleuca melaleuca
Merulius tremellosus
Mycena filopes
M. flavoalba
M. leptocephala
M. mucor
M. pearsoniana
M. polyadelpha
M. polygramma
M. pura
M. smithiana
M. speirea
M. stylobates
M. vitilis
Ripartites tricholoma
Russula nitida
Stereum hirsutum
S. rugosum
Tubaria furfuracea
Xerocomus badius
Xylaria longipes

Antrodia semisupina
Ascocoryne cylichnium
Cerocortitium confluens
Chondrostereum purpureum
Cylindrobasidium evolvens
Flagelloscypha minutissima
Hymenoscyphus immutabilis
H. rubicola
H. scutula sl.
Hyphoderma praetermissum
H. pallidum
Incrustoporia nivea
Kneiffiella sambuci
Mollisia spec.
Nectria spec. (imperf.)
Peniphora spec.
Perichaena corticalis
Pezizella alniella
Rutstroemia firma
Schizopora paradoxa
S. phellinoides
Sistotrema brinkmannii
Trechispora mutabilis
Typhula erythropus
T. phacorrhiza
T. sclerotioides
Zignoëlla ovoidea

c. Fungal species reported again, not reported again and newly found in plot 2.

reported again:

Armillariella cf. ostoyae
Collybia butyracea
C. dryphila
Crepidotus variabilis
Hypholoma fasciculare
Lactarius necator
Macrotyphula fistulosa
Mycena galericulata
M. galopoda
M. sanguinolenta
Oudemansiella platyphylla
Paxillus involutus
Psathyrella hydrophila
Stropharia aeruginosa
Xylaria hypoxylon

not reported again:

Amanita citrina
A. muscaria
L. subdulcis
L. vellereus

A. phalloides
A. vaginata
Calocybe ionides
Clitocybe brumalis
Collybia confluens
Coprinus disseminatus
Cortinarius alboviolaceus
C. multiformis
C. paleaceus
Hypholoma sublateritium
Laccaria amethystina
L. laccata
Lactarius blennius
L. chrysorrheus
L. vietus
Lycoperdon perlatum
Mycena corticola
M. epipterygia
Oudemansiella radicata
Russula delica
R. laurocerasi
R. nigricans
Scleroderma citrinum
Tricholoma sulphureum
Tricholomopsis rutilans
Xerocomus badius
X. subtomentosus

newly found:

Amanita rubescens
Bjerkandera adusta
Clitocybe candicans
C. diatreta
C. ditopa
C. gibba
C. metachroa
C. vibecina
Clitopilus hobsonii
Collybia cookei
C. peronata
Cortinarius sertipes
Entoloma juncinum
Galerina cinctula
G. hypnorum
Laccaria proxima
Lactarius quietus
L. theiogalus
Marasmiellus ramealis
Marasmius epiphylloides
Mycena cinerella
M. haematopoda
M. metata
M. polyadelpha
M. polygramma
M. vitilis
Phallus impudicus
Phlebia radiata
Piptoporus betulinus
Pleurotellus herbarum
Pluteus atricapillus
Polyporus brumalis
Psathyrella fulvescens
P. spadicea
P. spadiceogrisea
P. squamosa
Psilocybe crobula
Resupinatus trichotis
Russula emetica
R. ochroleuca
Stereum hirsitum
S. rugosum
Xerocomus chrysenteron
Ascocoryne spec. (imperf.)
Calocera cornea
Ceriporia excelsa
Cerocortitium confluens
Dacrymyces stillatus
Hymenoscyphus scutula sl.
Mollisia spec.
Nectria spec. (imperf.)
Peniphora quercina
Protocrea delicatula
Rutstroemia sydowiana
Schizopora paradoxa
Sphaerobolus stellatus
Trechispora farinacea
T. vaga
Typhula phacorrhiza

d. Fungal species report again, not reported again and newly found in plot 3.

reported again:

Armillariella ostoyae
Collybia butyracea
C. dryphila
Hypholoma fasciculare
Lactarius quietus
Mycena galericulata
M. galopoda
M. sanguinolenta
Oudemansiella platyphylla
Psathyrella hydrophila
Scleroderma citrinum
Stropharia aeruginosa
Xerocomus badius

not reported again:

Amanita citrina
A. phalloides
A. vaginata
Asterophora lycoperdoides
Calocybe ionides
Clitocybe brumalis
Collybia confluens
C. maculata
Coprinus disseminatus
Cortinarius elatior
C. hemitrichus
C. paleaceus
Dermocybe cinnabarina
Hygrocybe cantharellus
Hygrophoropsis aurantiaca
Hypholoma sublateritium
Inocybe geophylla
Laccaria amethystina
L. laccata
Lactarius camphoratus
L. chrysorrheus
L. mitissimus
L. subdulcis
L. vellereus
Marasmius androsaceus
M. rotula
Mycena epipterygia
M. polygramma
Oudemansiella radicata
Paxillus involutus
Russula cyanoxantha
R. delica
R. emetica
R. fragilis
R. nigricans
R. rubra
Xerocomus subtomentosus

newly found:

Amanita rubescens
Bjerkandera adusta
Clitocybe candicans
C. gibba
C. metachroa
C. vibecina
Clitopilus hobsonii
Collybia cookei
C. peronata
Galerina hypnorum
Gymnopilus penetrans
Inocybe napipes
Lactarius necator
L. theiogalus
Marasmiellus ramealis
Mycena cinerella
Stereum hirsutum
S. rugosum
Trametes versicolor
Xerocomus chrysenteron
Xylaria hypoxylon
Achroomyces peniophorae
Ascocoryne sarcoides
A. spec. (imperf.)
Botryobasidium subcoronatum
Calocera cornea
Cerocortitium confluens
Dacrymyces stillatus
Hyphoderma praetermissum
H. puberum
Mollisia spec.
Nectria spec. (imperf.)

M. inclinata	Peniphora quercina
M. rorida	P. violaceolivida
M. stylobates	Phanerochaete sordida
M. vitilis	Rutstroemia firma
Phallus impudicus	Schizopora paradoxa
Phlebia radiata	S. phellinoides
Piptoporus betulinus	Sphaerobolus stellatus
Pluteus atricapillus	Trametes spec. (juv.)
Psathyrella fulvescens	Trechispora cohaerens
Psilocybe crobula	T. farinacea
Resupinatus applicatus	Xenasma filicina
Russula ochroleuca	Zignoëlla ovoidea

Appendix 4D.
Fungi collected in Hackfort ('Oude Maat') in 1982-1984 by A.E. Jansen (J) and B.W.L. de Vries (V), conserved in the herbarium of the Biological Station at Wijster.

Agaricales

Agrocybe firma	J783 J856
Armillariella ostoyae	J803
Conocybe macrocephala	J842 J893 J894
Cortinarius alnetorum	J804
C. sertipes	V4571 V4826 J805 J858 J891
Entoloma juncinum	V4828
Marasmius epiphylloides	J844
Melanoleuca melaleuca	J901
Mycena flavoalba	J889
M. pearsoniana	J892
M. polyadelpha	J855
Mycenella margaritispora	J887
Naucoria escharoides	V4596 J895
Oudemansiella badia	J888
O. radicata	J890
Pleurotellus herbarum	J859 J896
Pluteus thomsonii	J843
Psathyrella hydrophila	V4973
P. spadicea	J806
Resupinatus trichotis	V4606

Aphyllophorales + Gasteromycetes

Achroomyces peniophorae	V4602
Ceriporia excelsa	V5064
Clavariadelphus junceus	V4975
Flagelloscypha minutissima	V4999
Hyphoderma praetermissum	V5010

Incrustoporia nivea	J848 J902
Kneifiella sambuci (comb.prov.)	V5021
Macrotyphula fistulosa	V4976
Peniophora quercina	V5040 V5039
P. violaceolivida	V5041
Piptoporus betulinus	V4935
Schizopora phelliniodes	V4974 (V4608?)
Stereum hirsutum	J850
Trechispora cohaerens	V5057
T. mutabilis	V5058
Typhula phacorrhiza	V4977
T. sclerotioides	V845 J809
Xenasma filicinum	V5063

Ascomyceten

Ascocoryne cylichnium	J808
Hymenoscyphus immutabilis	J782 J846
H. rubicola	J857
Nectria (imperfect)	V5034 J849
Pezizella alniella	J807
Protocrea delicatula	V4603
Xylaria longipes	J847

Myxomyceten

Perichaena corticalis	V5042

Appendix 5. Contour plots of solute concentrations in soil solution as a function of time and depth.

Each page shows contour plots for each of the four plots A, B, C and D. The contour plots were calculated and drawn by the programme Isopas, written by G. Oerlemans. Units are μS/cm for the electrical conductivity, mmol/l for organic C and for SiO_2. For ionic species, equivalent ionic concentrations in mmol/l (formally meq/l) are used. The degree of shading shows the value for the parameter in question from the lowest range (blank) to the highest range (black) as a function of depth below the soil surface (vertical scale) and time between March 1981 and May 1984 (horizontal scale). The values for the contour intervals between white and black are shown at the top of each set of four plots. Contour intervals were chosen so that changes in concentrations with time and depth were expressed most clearly. The plots were derived by means of a quadratic interpolation between the value of each measured point in time and space, and of the 10 closest neighbouring points. Missing values were interpolated linearly (over time only) between adjacent points, and duplicate data (generally at 10, 40 and 90 cm depth) were averaged. The diagram on the next page shows the positions of the actual data points on the depth-time coordinates used.

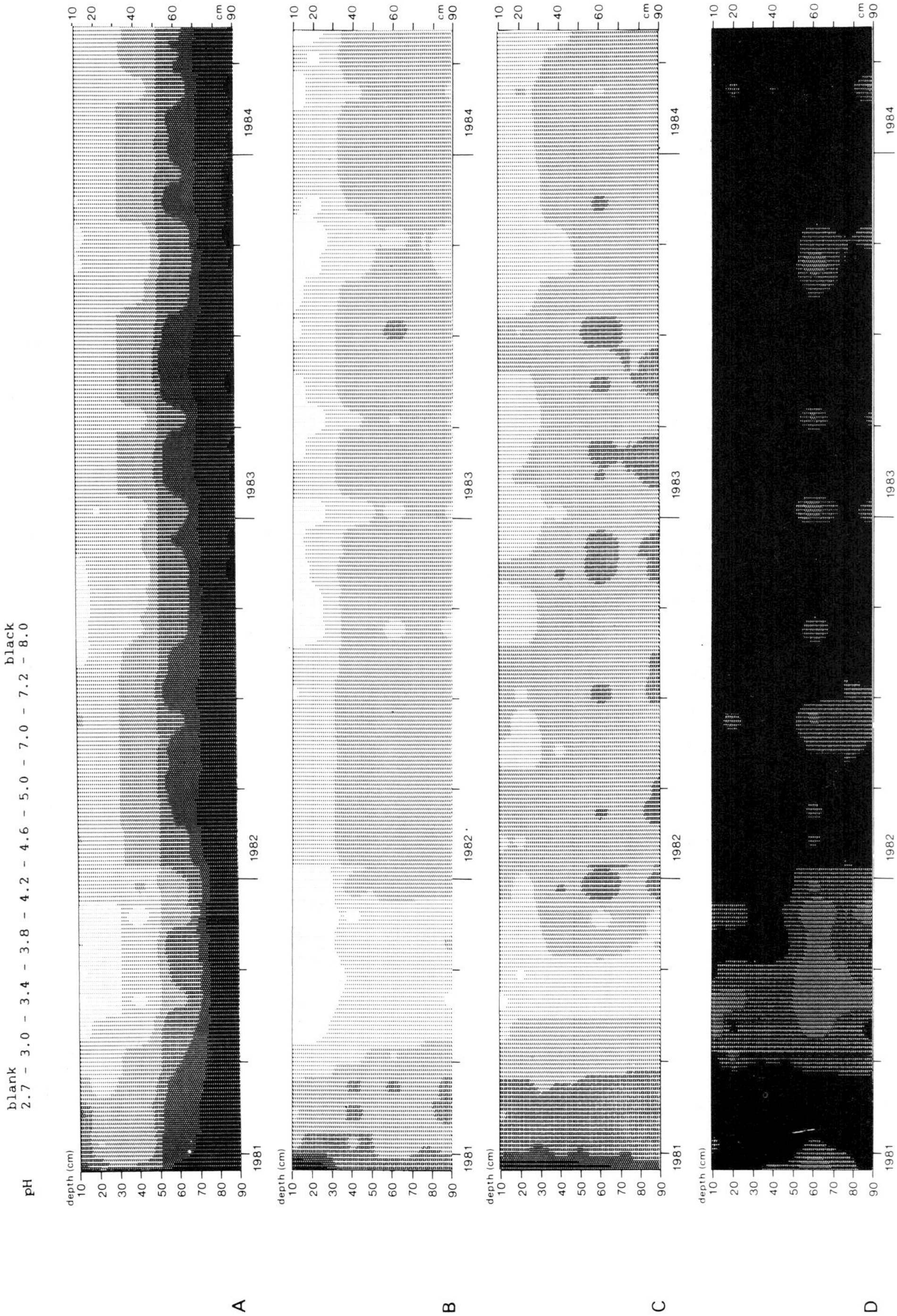
pH
blank 2.7 - 3.0 - 3.4 - 3.8 - 4.2 - 4.6 - 5.0 - 7.0 - 7.2 - 8.0 black
depth (cm)
1981
1982
1983
1984
A
B
C
D

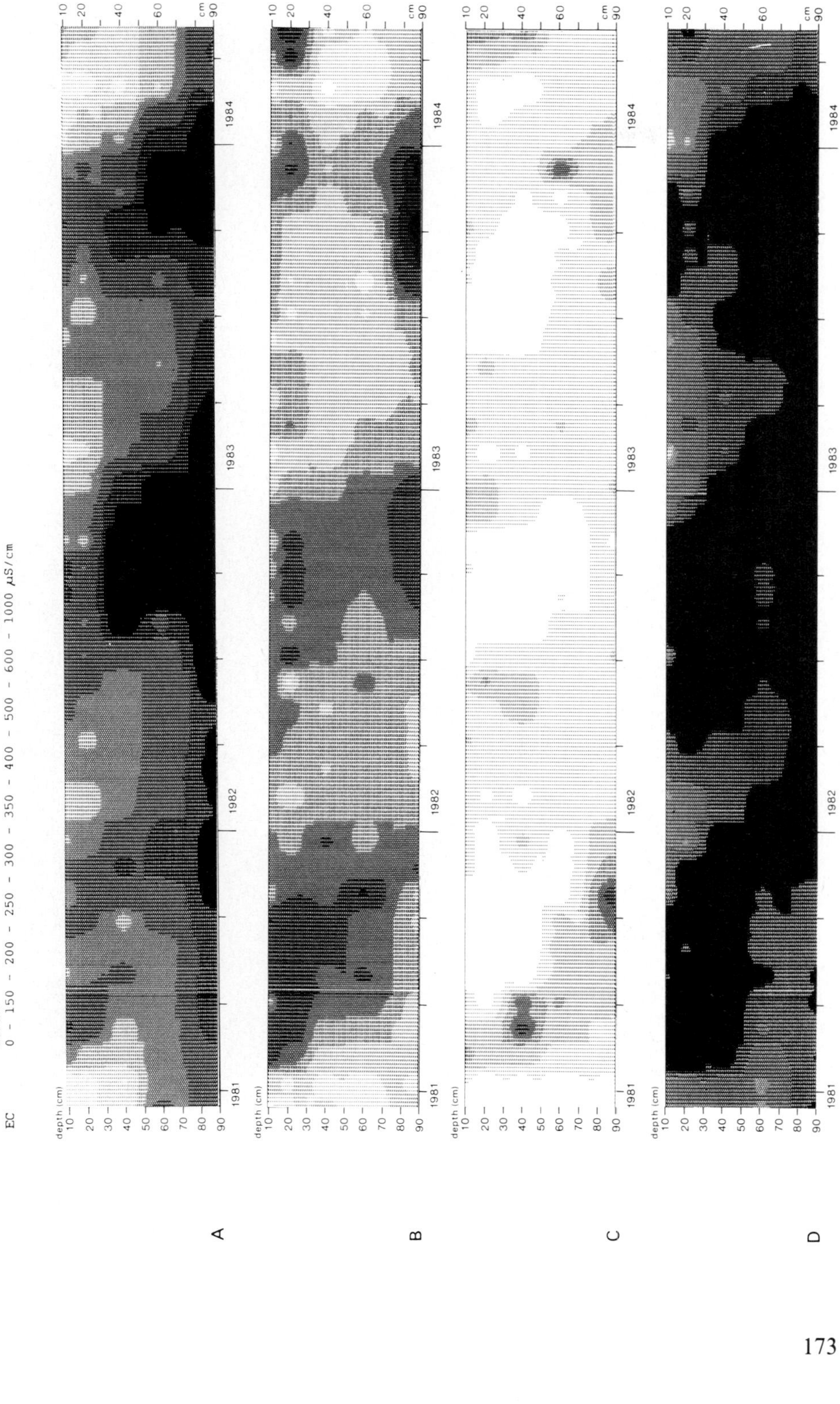
EC
blank
black
0 - 150 - 200 - 250 - 300 - 350 - 400 - 500 - 600 - 1000 µS/cm
depth (cm)
1981
1982
1983
1984
A
B
C
D

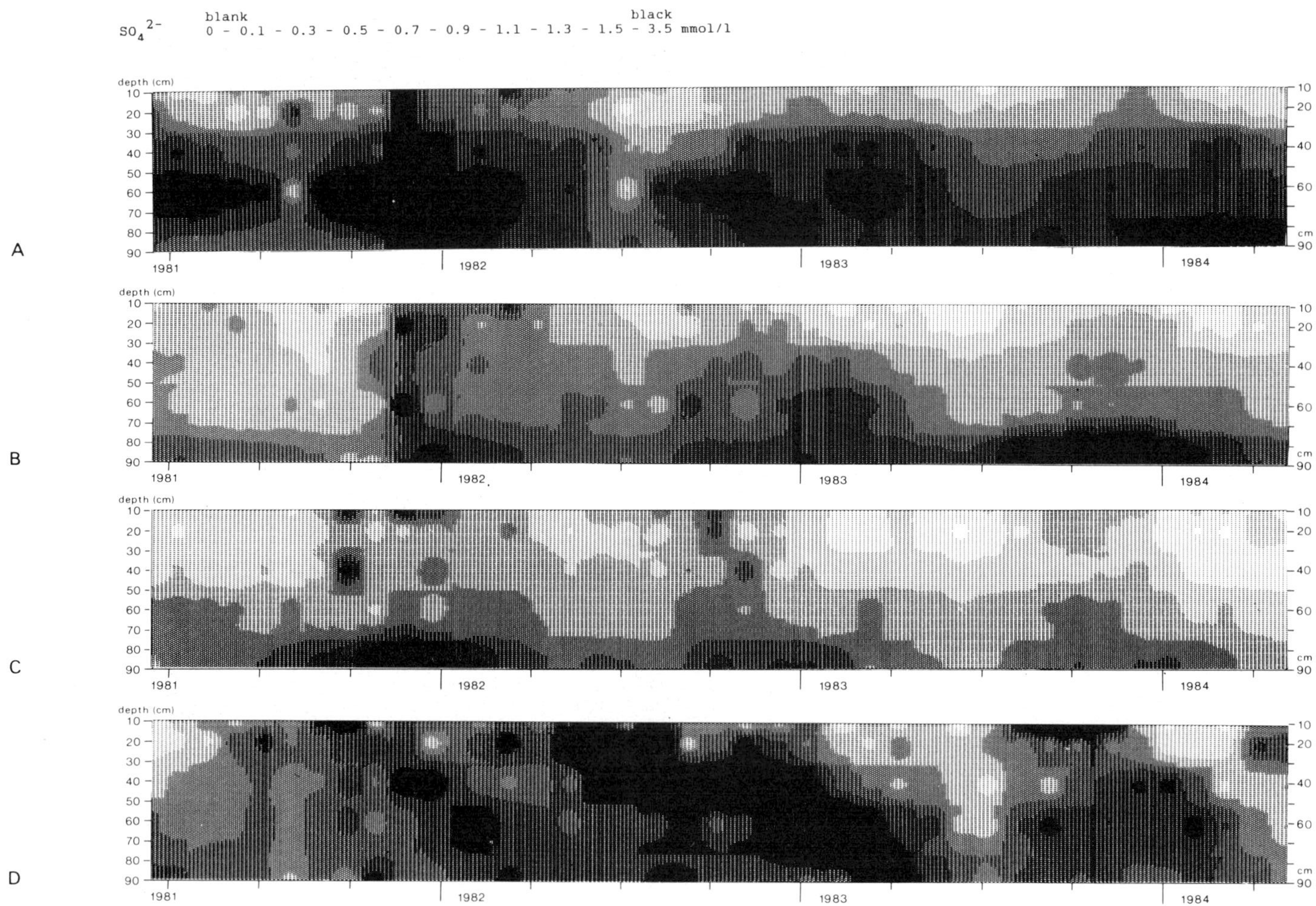

SO_4^{2-}
blank
black
0 - 0.1 - 0.3 - 0.5 - 0.7 - 0.9 - 1.1 - 1.3 - 1.5 - 3.5 mmol/l
A
B
C
D
depth (cm)
10
20
30
40
50
60
70
80
90
cm
1981
1982
1983
1984

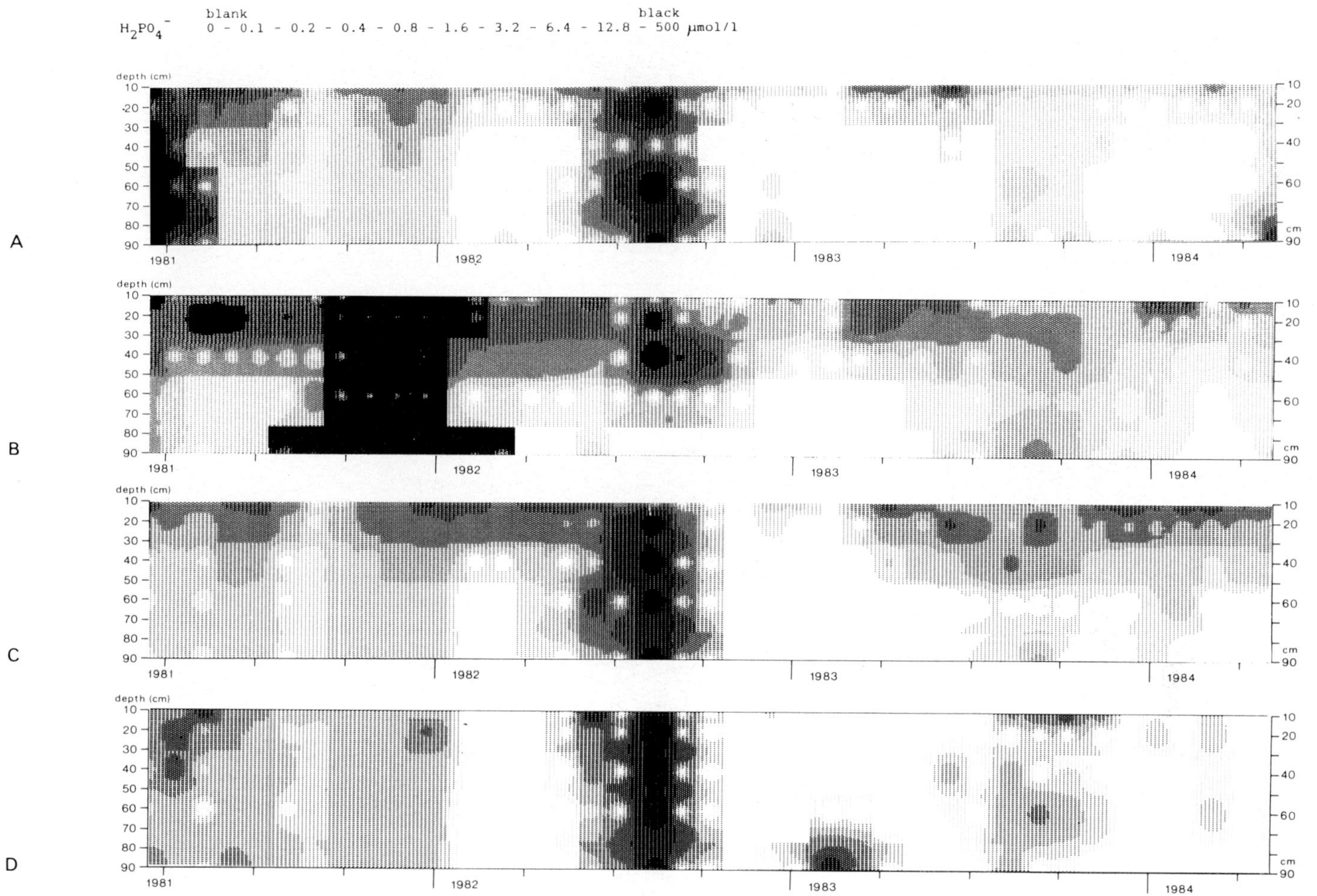

$H_2PO_4^-$
blank
black
0 - 0.1 - 0.2 - 0.4 - 0.8 - 1.6 - 3.2 - 6.4 - 12.8 - 500 μmol/l
A
B
C
D
depth (cm)
10
20
30
40
50
60
70
80
90
cm
1981
1982
1983
1984

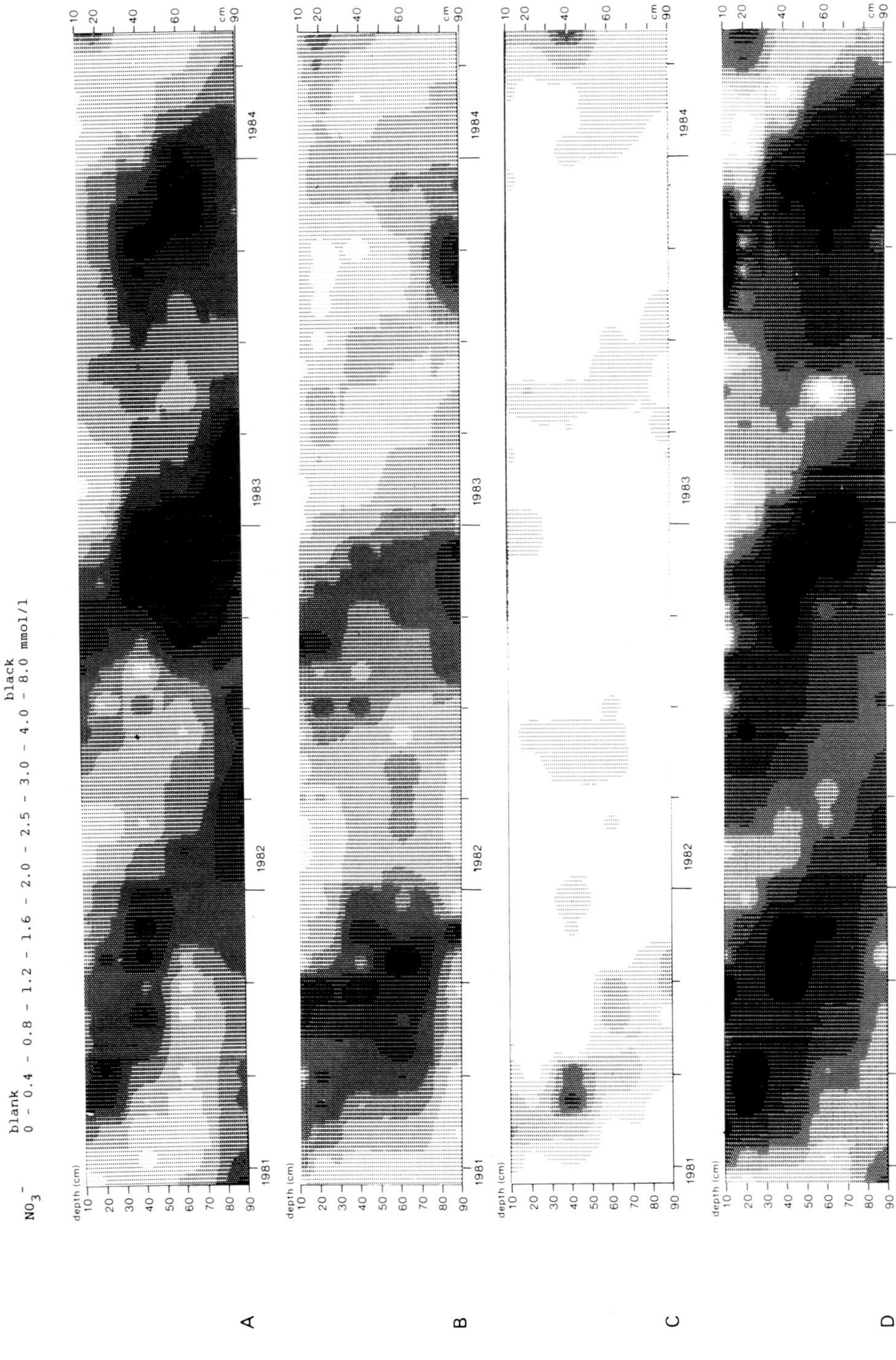
NO_3^-
blank
0 - 0.4 - 0.8 - 1.2 - 1.6 - 2.0 - 2.5 - 3.0 - 4.0 - 8.0 mmol/l
black
depth (cm)
10 20 30 40 50 60 70 80 90
cm
1981
1982
1983
1984
A
B
C
D

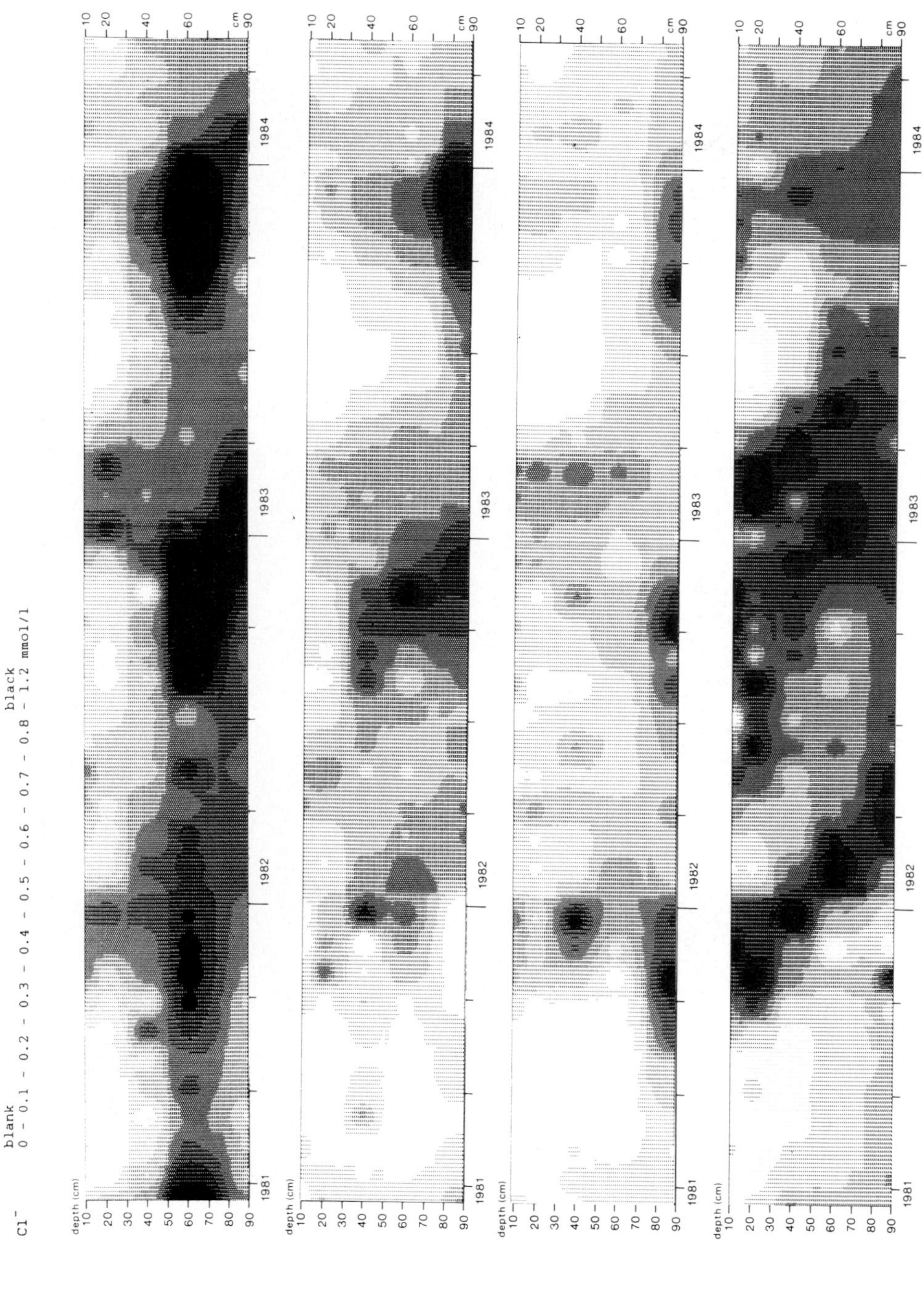

Cl⁻
blank
black
0 - 0.1 - 0.2 - 0.3 - 0.4 - 0.5 - 0.6 - 0.7 - 0.8 - 1.2 mmol/l
depth (cm)
10
20
30
40
50
60
70
80
90
cm
1981
1982
1983
1984
A
B
C
D

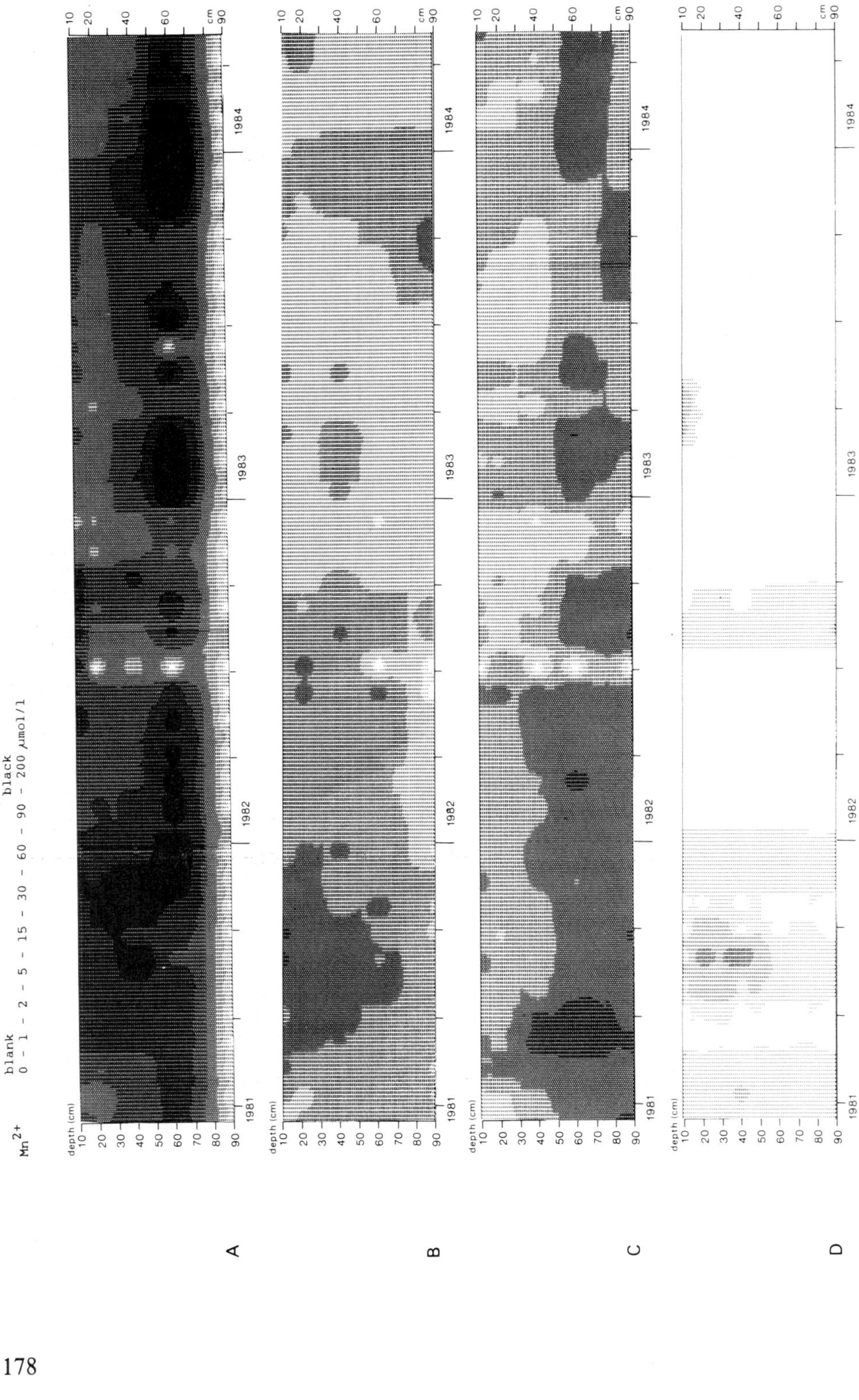
Mn2+
blank
0 - 1 - 2 - 5 - 15 - 30 - 60 - 90 - 200 µmol/l
black
depth (cm)
1981
1982
1983
1984
A
B
C
D

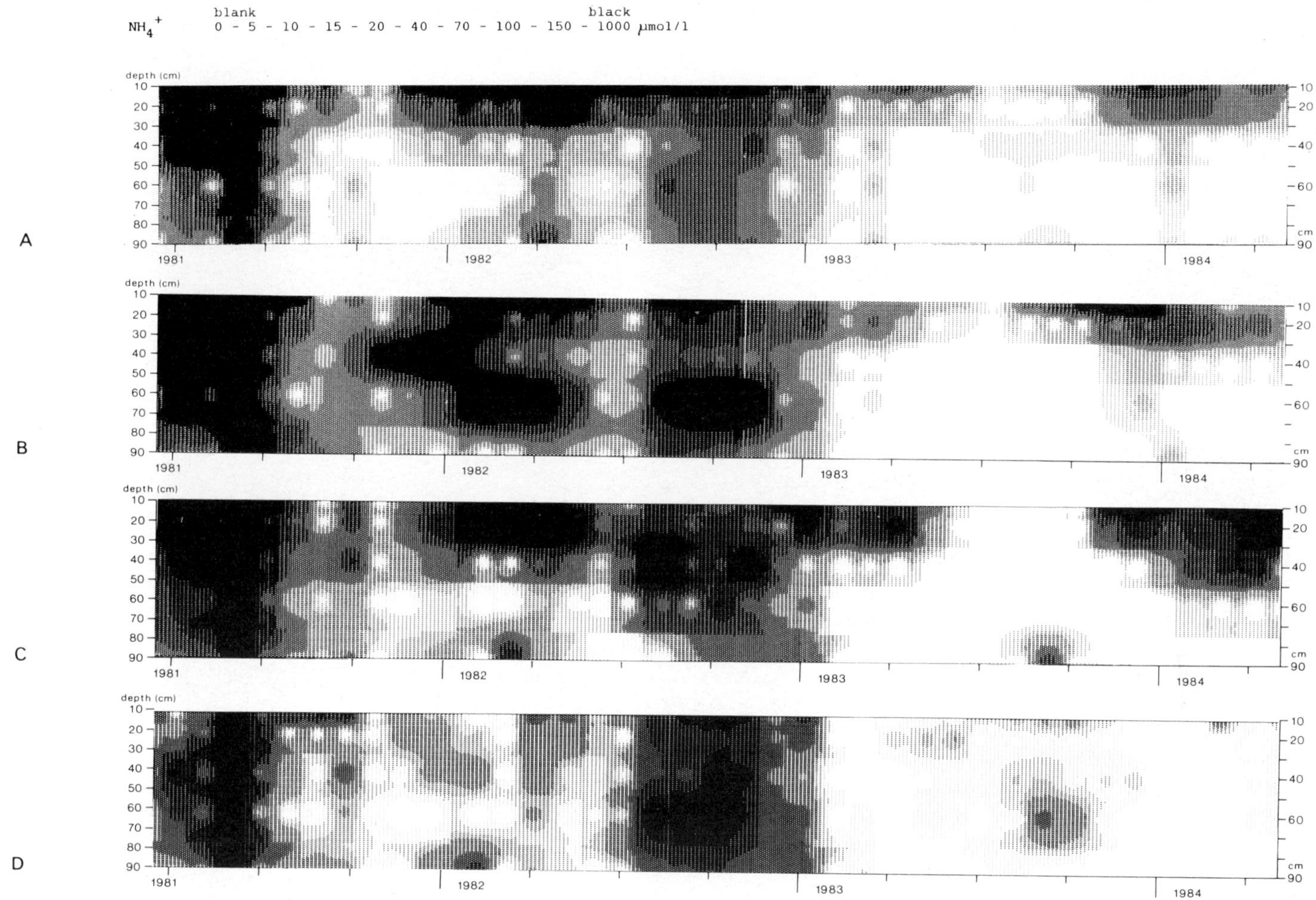
NH_4^+
blank
black
0 - 5 - 10 - 15 - 20 - 40 - 70 - 100 - 150 - 1000 µmol/l
depth (cm)
10
20
30
40
50
60
70
80
90
cm
1981
1982
1983
1984
A
B
C
D

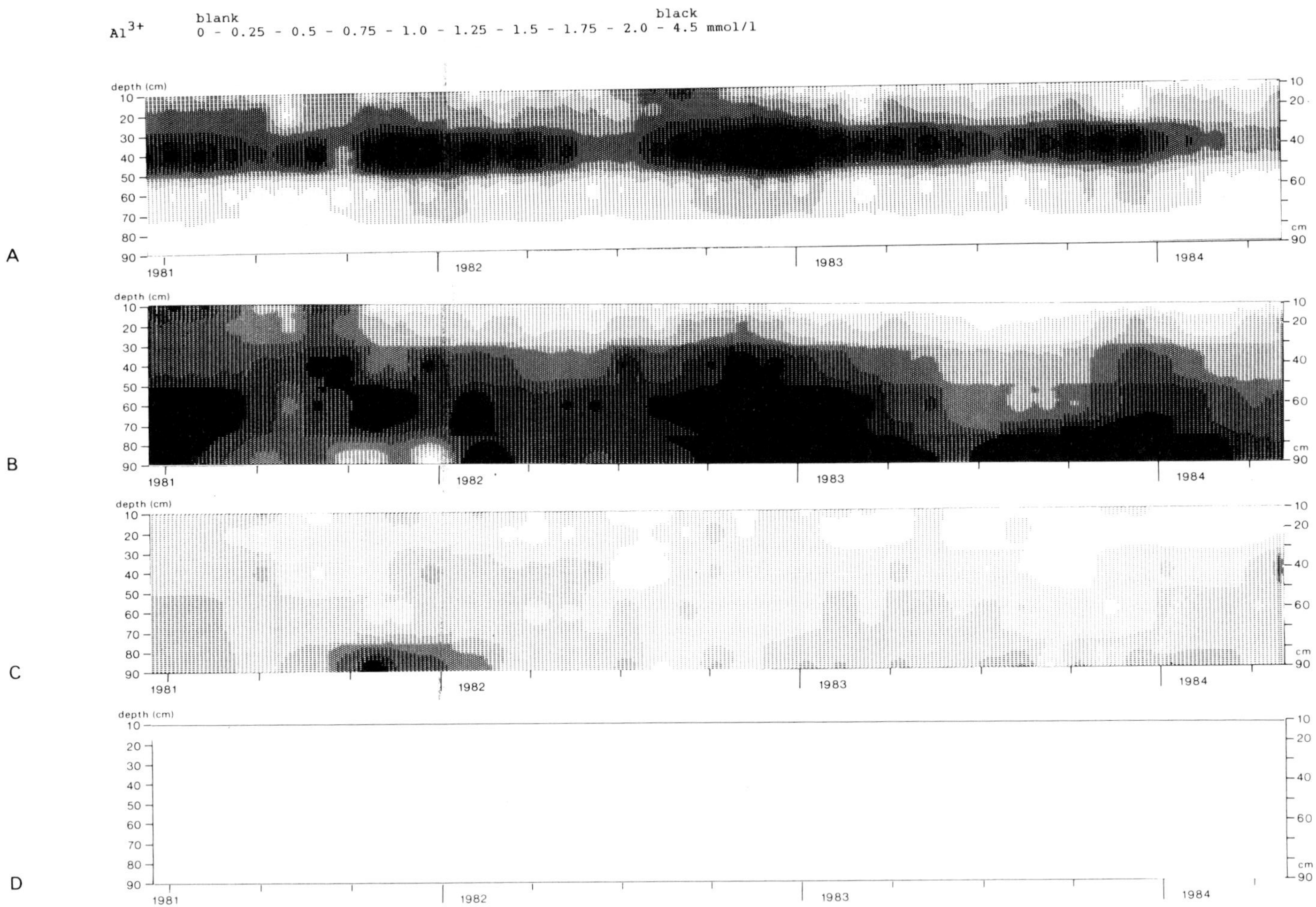

Al^{3+}
blank
black
0 - 0.25 - 0.5 - 0.75 - 1.0 - 1.25 - 1.5 - 1.75 - 2.0 - 4.5 mmol/l
A
B
C
D
depth (cm)
cm
1981
1982
1983
1984

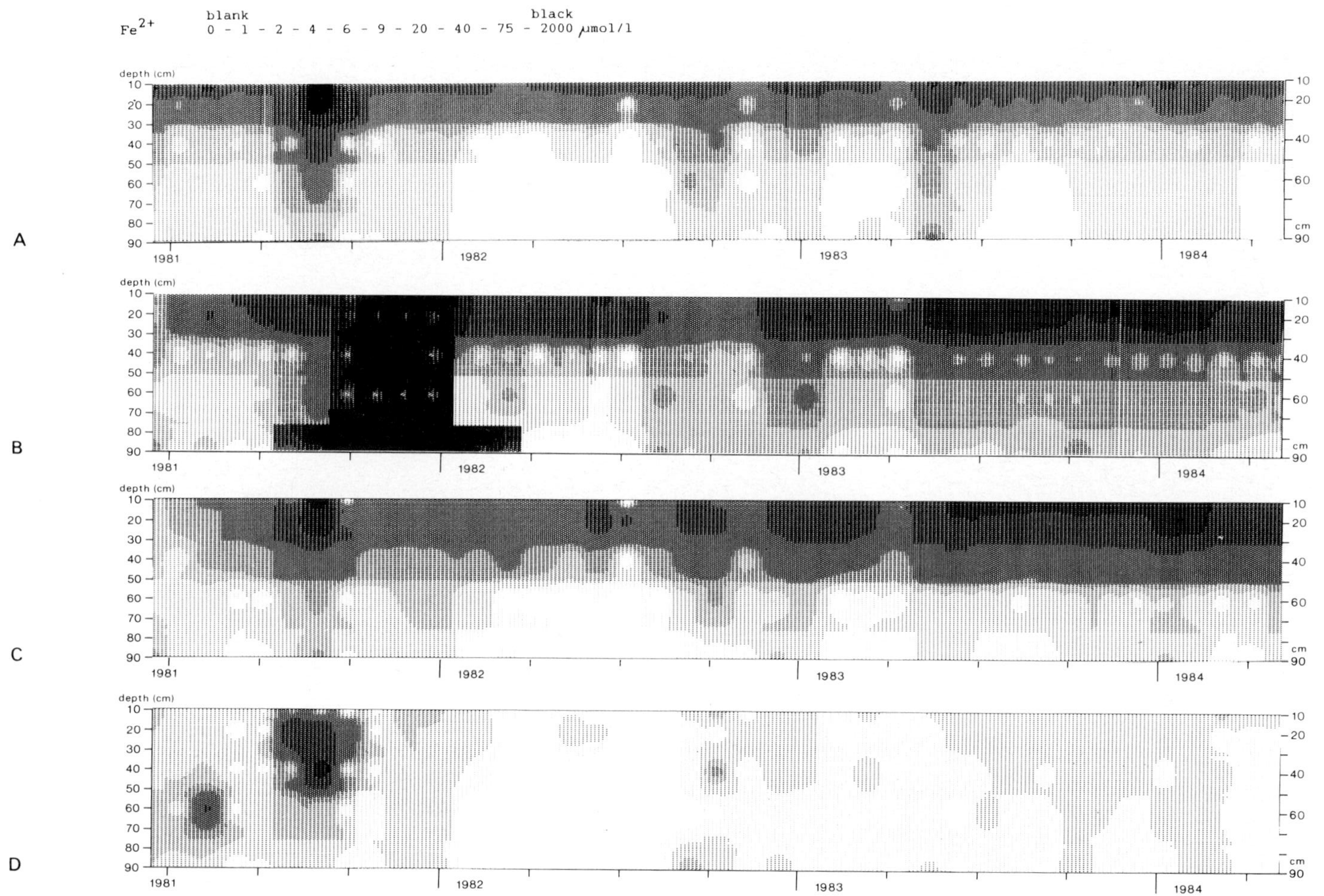

Fe^{2+}
blank
black
0 - 1 - 2 - 4 - 6 - 9 - 20 - 40 - 75 - 2000 μmol/l
depth (cm)
10
20
30
40
50
60
70
80
90
cm
1981
1982
1983
1984
A
B
C
D

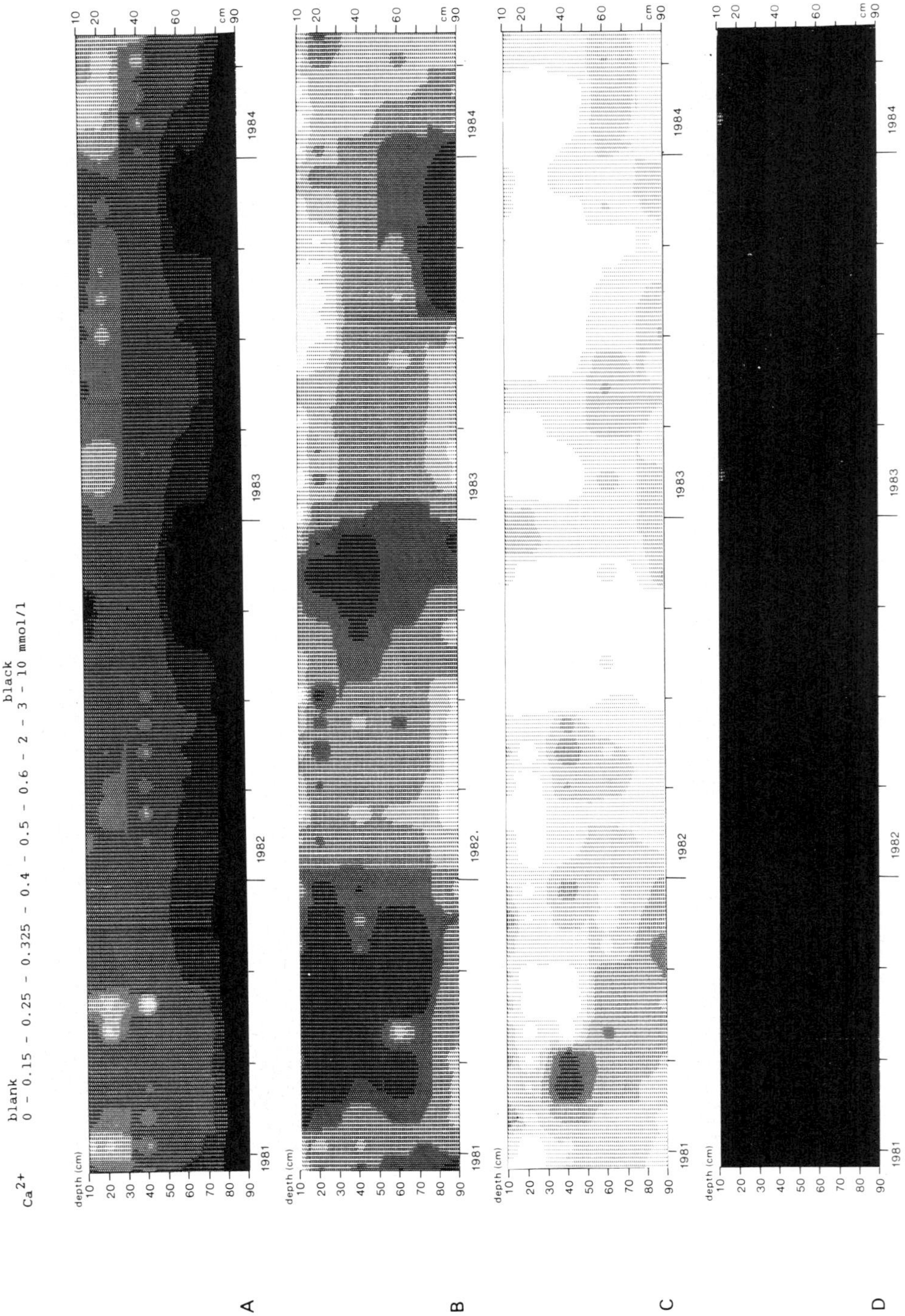
Ca^{2+}
blank
black
0 - 0.15 - 0.25 - 0.325 - 0.4 - 0.5 - 0.6 - 2 - 3 - 10 mmol/l
depth (cm)
cm
1981
1982
1983
1984
A
B
C
D

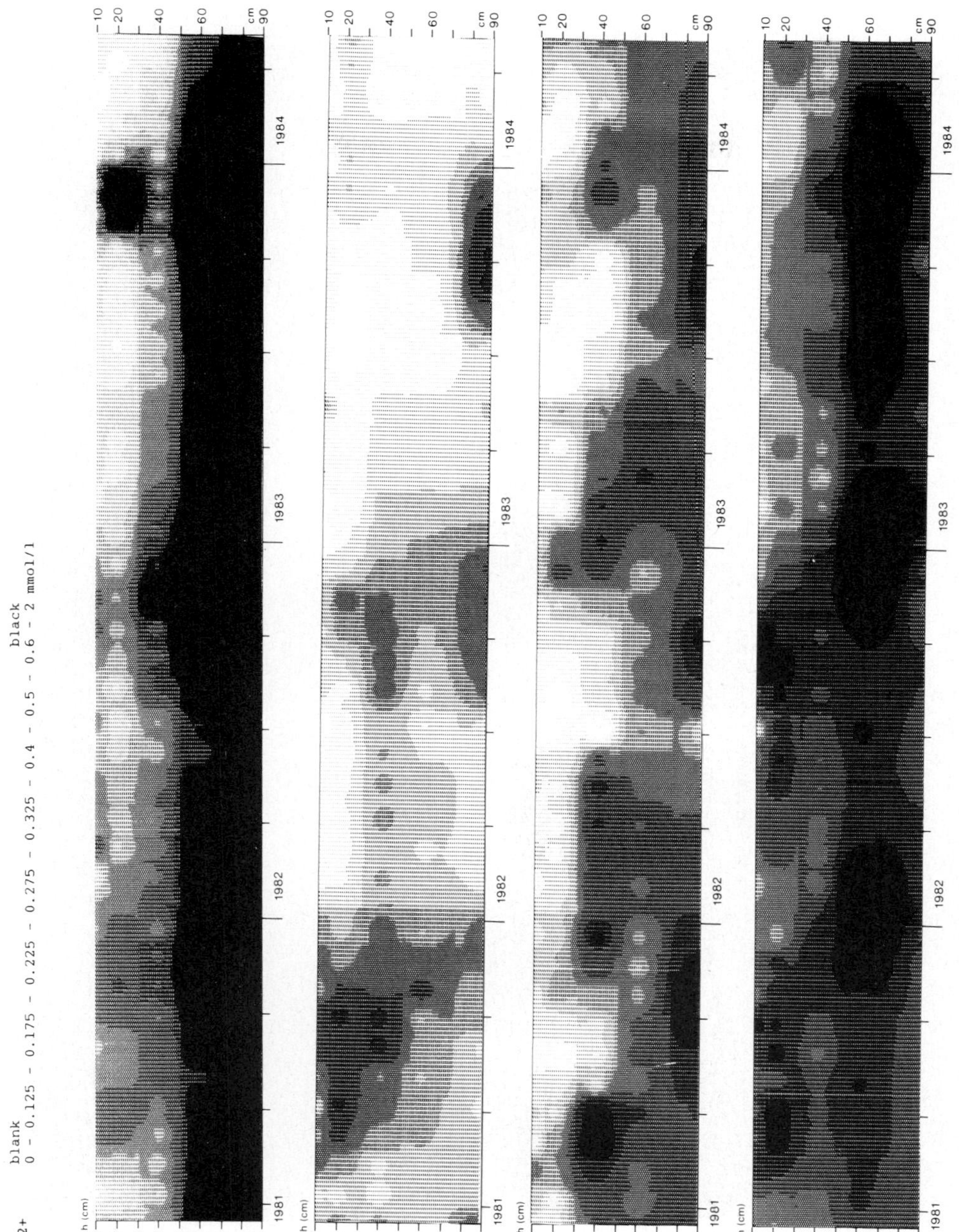
Mg2+
blank
black
0 - 0.125 - 0.175 - 0.225 - 0.275 - 0.325 - 0.4 - 0.5 - 0.6 - 2 mmol/l
depth (cm)
cm
1981
1982
1983
1984
A
B
C
D

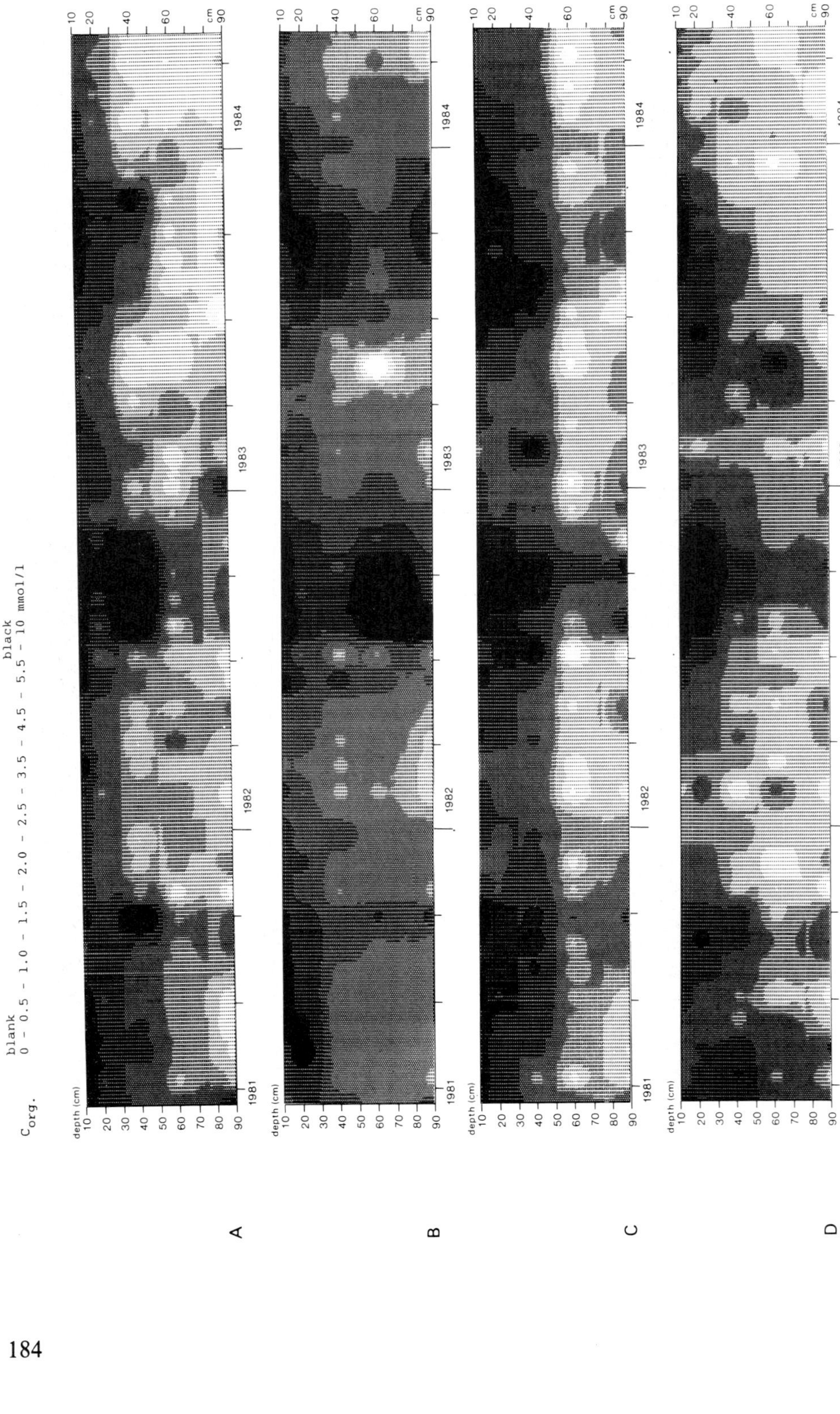
C org.
blank
0 - 0.5 - 1.0 - 1.5 - 2.0 - 2.5 - 3.5 - 4.5 - 5.5 - 10 mmol/l
black
depth (cm)
10 20 30 40 50 60 70 80 90
cm
1981
1982
1983
1984
A
B
C
D

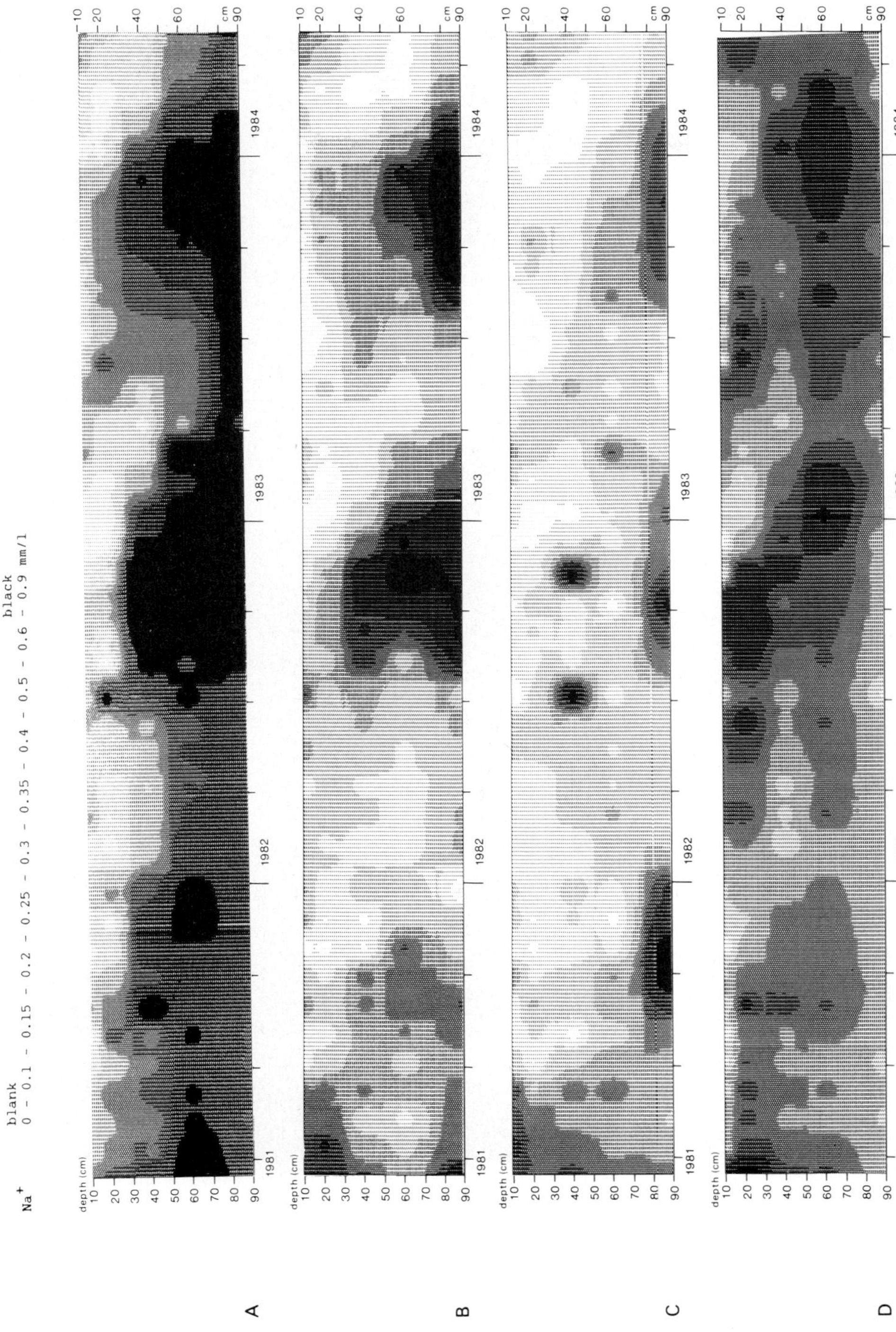

Na+
blank
0 - 0.1 - 0.15 - 0.2 - 0.25 - 0.3 - 0.35 - 0.4 - 0.5 - 0.6 - 0.9 mm/l
black
A
B
C
D
depth (cm)
10 20 30 40 50 60 70 80 90
cm
1981
1982
1983
1984

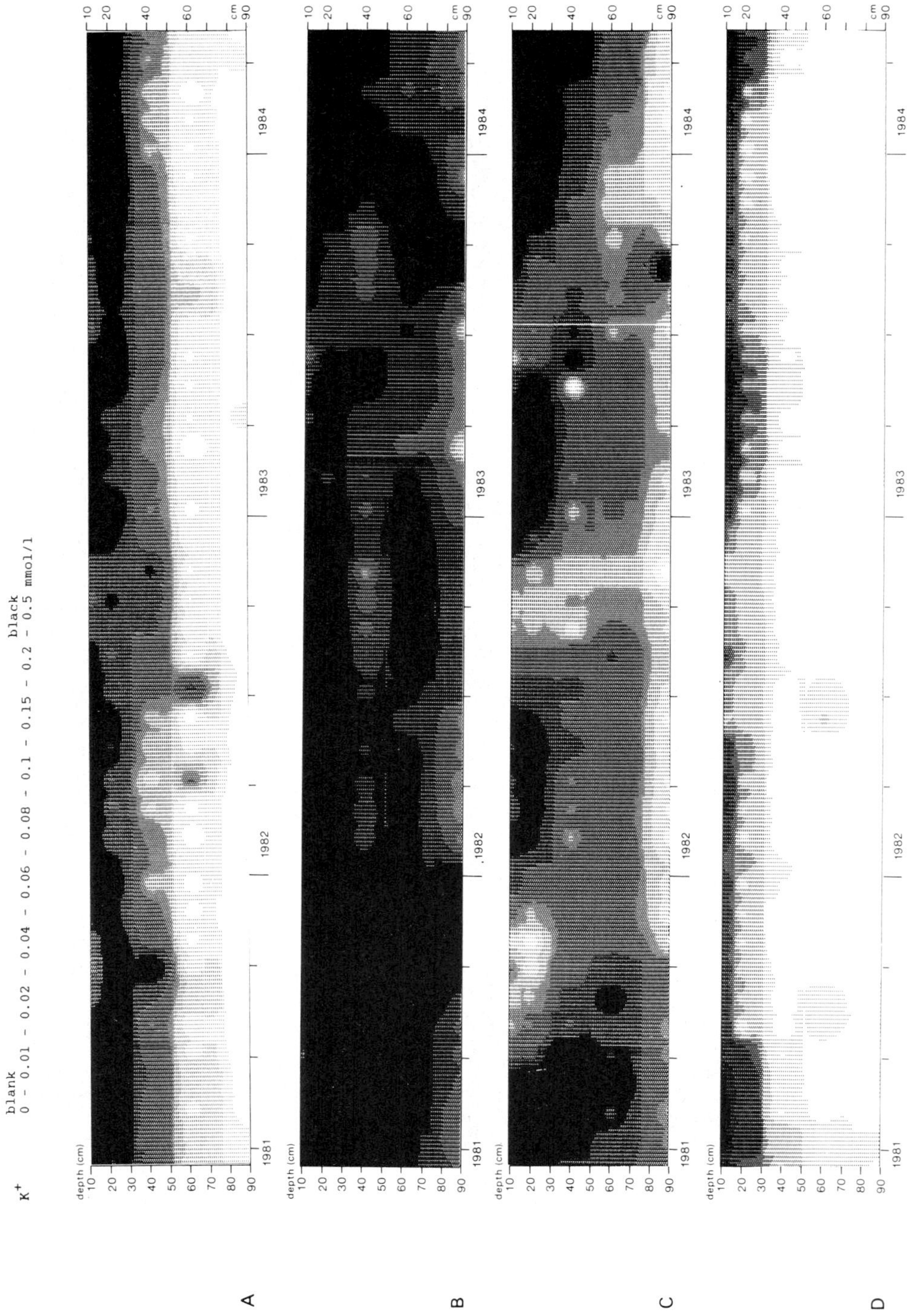
K+
blank
0 - 0.01 - 0.02 - 0.04 - 0.06 - 0.08 - 0.1 - 0.15 - 0.2 - 0.5 mmol/l
black
depth (cm)
10
20
30
40
50
60
70
80
90
cm
1981
1982
1983
1984
A
B
C
D

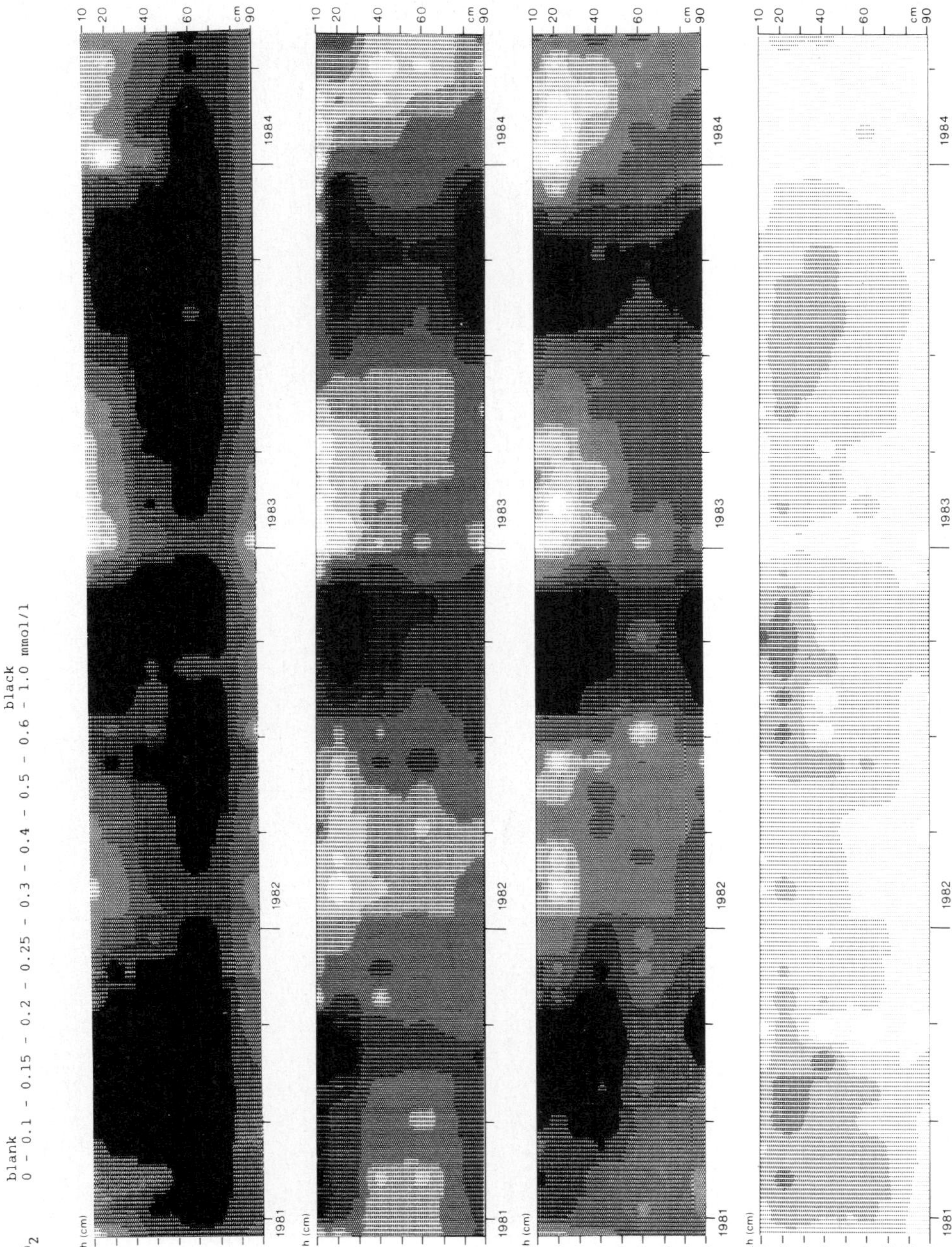
SiO2
blank
0 - 0.1 - 0.15 - 0.2 - 0.25 - 0.3 - 0.4 - 0.5 - 0.6 - 1.0 mmol/l
black
depth (cm)
10 20 30 40 50 60 70 80 90
cm
1981
1982
1983
1984
A
B
C
D

Appendix 6. Above-ground and below-ground monthly waterfluxes in three hydrological years.

Above-ground and below-ground monthly water fluxes (in cm) at each of the plots A, B, C and D, in three hydrological years: 4.1.1981 – 3.22.1982, 3.22.1982 – 3.22.1983 and 3.22.1983 – 3.19.1984. Time refers to days after the start of the hydrological year. Infiltration was assumed to be equal to the measured throughfall flux, the other fluxes were derived from SWATRE simulations. At the soil depths for which fluxes were calculated, soil solutions were sampled.

PLOT A

	Time (days)	Trans-piration	Infil-tration	depth(cm) 10	20	40	60	90
1981-1982	28.	0.000	0.770	0.105	-0.068	-0.734	-3.764	-6.449
	56.	2.172	4.580	-3.063	-1.995	-0.709	-0.334	-1.403
	84.	6.212	2.680	-2.269	-2.011	-1.209	0.287	-0.444
	112.	6.409	9.720	-9.975	-7.599	-4.333	-0.813	0.867
	140.	7.321	2.750	-2.792	-1.975	-0.687	0.756	-0.882
	168.	5.405	2.280	-0.942	-0.246	0.386	1.592	1.190
	196.	1.884	6.530	-5.540	-2.798	0.433	1.575	2.120
	224.	0.017	3.900	-4.130	-4.867	-4.457	-2.738	0.436
	252.	0.000	6.540	-6.141	-5.516	-3.639	-0.771	0.232
	291.	0.000	4.740	-5.521	-5.334	-6.611	-7.734	-7.734
	328.	0.000	2.430	-2.255	-2.317	-2.451	-4.527	-5.426
	357.	0.000	4.520	-3.412	-2.859	-2.409	-1.733	-1.077
	Total	29.419	51.440	-45.935	-37.584	-26.422	-18.204	-18.568
1982-1983	32.	0.000	1.290	-0.234	-0.205	-0.380	-1.140	-4.560
	63.	2.279	3.600	-2.413	-1.649	-0.896	-0.374	-0.499
	91.	6.559	5.660	-4.923	-2.930	0.056	1.680	1.304
	123.	9.064	4.490	-4.807	-4.520	-3.805	-1.159	-1.884
	154.	6.213	6.120	-4.151	-1.644	0.120	2.014	2.074
	182.	5.076	1.260	-1.094	-1.235	0.036	1.816	1.459
	211.	1.759	6.580	-5.298	-3.967	-0.473	1.525	2.250
	243.	0.380	5.820	-5.401	4.657	-4.204	-2.433	-1.114
	273.	0.000	5.730	-5.784	-5.416	-5.326	-3.512	-0.293
	305.	0.342	6.380	-6.889	-6.244	-5.822	-5.490	-5.490
	337.	0.000	3.580	-3.841	-3.723	-4.120	-6.687	-6.827
	365.	0.000	1.330	-0.431	-0.450	-0.753	-1.077	-1.558
	Total	31.672	51.930	-45.267	-36.639	-25.566	-14.836	-15.139
1983-1984	34.	0.000	6.430	-4.669	-4.223	-3.390	-2.882	-2.729
	65.	0.862	9.480	-8.260	-7.236	-5.479	-3.394	-3.519
	91.	6.567	2.950	-2.807	-2.577	-2.397	-3.288	-5.601
	119.	9.745	3.440	-3.355	-2.084	-0.255	1.766	-0.405
	154.	7.382	4.530	-1.328	-0.034	0.129	2.428	1.854
	182.	3.477	4.860	-5.012	-2.567	0.033	1.662	2.128
	214.	1.835	2.210	-1.538	-1.480	-0.027	1.812	1.818
	245.	0.000	0.210	-0.030	-0.338	-0.452	0.077	0.182
	273.	0.000	7.870	-8.554	-7.154	-5.957	-4.801	-0.950
	301.	0.000	10.290	-10.243	-8.384	-5.916	-3.192	-2.530
	330.	0.000	7.340	-7.740	-7.820	-7.820	-7.820	-7.820
	361.	0.000	2.350	-1.606	-1.812	-4.391	-6.602	-6.602
	Total	29.868	61.960	-55.142	-45.709	-35.922	-24.234	-24.174

PLOT B

	Time (days)	Transpiration	Infiltration	depth(cm) 10	20	40	60	90
1981-1982	28.	0.000	0.800	0.509	0.337	-0.194	-1.201	-5.725
	56.	2.110	4.630	-3.502	-2.651	-1.530	-0.874	-1.283
	84.	6.290	2.690	-2.233	-1.742	-1.061	-0.027	-0.639
	112.	6.901	9.580	-9.231	-6.991	-3.765	-0.185	0.846
	140.	7.330	2.560	-2.419	-1.610	-0.620	0.527	-0.406
	168.	5.065	2.210	-1.073	-0.422	0.143	0.811	0.358
	196.	1.672	6.450	-5.523	-3.035	-0.126	0.667	0.661
	224.	0.020	3.700	-3.781	-4.290	-3.832	-0.980	1.352
	252.	0.000	6.840	-6.647	-5.953	-5.077	-2.308	2.010
	291.	0.000	4.650	-5.245	-5.231	-5.578	-6.948	-8.634
	328.	0.000	2.590	-2.357	-2.452	-2.818	-3.331	-6.142
	357.	0.000	4.970	-4.047	-3.601	-3.165	-2.781	-1.241
	Total	29.389	51.670	-45.548	-37.642	-27.623	-16.629	-18.844
1982-1983	32.	0.000	1.470	-0.236	-0.231	-0.315	-0.570	-2.360
	63.	2.280	3.650	-2.700	-2.126	-1.452	-0.970	-1.106
	91.	6.698	5.700	-4.854	-3.117	-0.777	0.851	0.122
	123.	9.257	4.350	-4.287	-3.763	-2.665	-0.347	-0.256
	154.	5.529	5.940	-4.562	-2.452	0.005	0.726	0.352
	182.	4.115	0.880	-0.723	-0.726	-0.294	0.526	0.503
	211.	0.985	6.090	-4.832	-3.407	-0.431	0.570	0.519
	243.	0.130	5.130	-4.843	-4.163	-3.503	-0.295	0.802
	273.	0.000	5.820	-5.898	-5.431	-5.098	-4.588	-0.142
	305.	0.740	6.640	-7.079	-6.603	-6.213	-5.187	-4.779
	337.	0.000	4.080	-4.085	-4.265	-4.453	-4.709	-7.929
	365.	0.000	1.380	-0.647	-0.585	-0.797	-1.205	-2.313
	Total	29.735	51.310	-44.697	-36.870	-25.994	-15.199	-16.589
1983-1984	34.	0.000	6.770	-5.178	-4.793	-4.340	-3.316	-2.210
	65.	0.830	9.100	-7.850	-6.689	-5.555	-4.868	-3.637
	91.	7.280	3.080	-2.907	-2.576	-2.073	-1.084	-4.705
	119.	9.481	3.540	-3.104	-2.051	-0.626	1.080	0.059
	154.	5.399	4.140	-1.966	-0.208	0.026	0.846	0.352
	182.	2.858	4.720	-4.425	-2.831	-0.014	0.425	0.380
	214.	1.567	1.800	-1.156	-1.172	-0.698	0.690	0.781
	245.	0.000	0.270	0.074	-0.127	-0.138	0.615	0.654
	273.	0.000	9.050	-9.467	-7.923	-6.014	-4.078	-1.037
	301.	0.000	10.980	-10.941	-9.945	-7.834	-4.254	-0.317
	330.	0.000	6.710	-7.012	-6.525	-7.037	-7.037	-7.037
	361.	0.000	2.210	-1.696	-1.920	-3.338	-6.672	-9.709
	Total	27.415	62.370	-55.627	-46.759	-37.640	-27.653	-26.426

PLOT C

	Time (days)	Transpiration	Infiltration	depth(cm) 10	20	40	60	90
1981-1982	28.	0.010	0.810	0.356	0.041	-2.110	-7.639	-13.132
	56.	2.120	4.600	-3.490	-2.676	-1.529	-0.921	-3.211
	84.	6.240	2.640	-2.271	-1.683	-1.004	0.234	-0.406
	112.	6.672	9.330	-9.332	-7.502	-4.070	-0.681	1.421
	140.	7.295	2.600	-2.675	-1.710	-0.459	1.252	-0.397
	168.	3.572	2.340	-1.483	-0.159	0.198	1.252	0.920
	196.	1.549	6.500	-5.817	-3.776	-0.508	0.994	1.286
	224.	0.020	3.720	-3.875	-4.275	-5.510	-4.312	1.351
	252.	0.000	6.620	-6.218	-5.633	-0.910	4.488	7.345
	291.	0.000	4.870	-5.299	-5.430	-9.183	-10.868	-10.868
	328.	0.000	2.560	-2.393	-2.507	-2.813	-6.474	-8.512
	357.	0.000	5.170	-4.232	-3.847	-3.469	-2.252	-0.525
	Total	27.474	51.760	-46.728	-39.157	-31.007	-24.927	-24.726
1982-1983	32.	0.000	1.420	-0.085	-0.136	-0.450	-1.603	-8.880
	63.	2.300	3.660	-2.739	-2.091	-1.374	-0.839	-1.221
	91.	6.576	5.570	-4.950	-3.208	-0.362	1.617	1.226
	123.	8.213	4.420	-4.732	-4.141	-3.473	-1.028	-1.168
	154.	4.205	6.210	-4.668	-2.753	-0.068	1.099	1.041
	182.	4.791	0.920	-0.805	-0.670	-0.486	1.124	0.953
	211.	0.935	6.230	-5.066	-3.772	-1.769	0.357	0.998
	243.	0.050	5.320	-5.070	-4.671	-3.948	-3.543	-2.467
	273.	0.000	6.040	-6.158	-5.771	-5.447	-2.355	4.964
	305.	0.636	6.750	-6.819	-6.413	-5.884	-4.072	-4.072
	337.	0.000	4.030	-3.900	-3.907	-4.214	-8.899	-8.899
	365.	0.000	1.370	-0.538	-0.484	-0.681	-1.375	-2.990
	Total	27.706	51.940	-45.530	-38.015	-28.156	-19.515	-20.517
1983-1984	34.	0.000	6.740	-5.009	-4.848	-4.233	-3.167	-2.295
	65.	0.720	8.450	-7.037	-6.240	-5.232	-2.159	-2.212
	91.	6.925	3.090	-2.949	-2.419	-1.279	-3.773	-8.607
	119.	8.827	3.560	-3.417	-2.151	-0.573	1.537	-1.992
	154.	4.126	4.200	-1.748	-0.150	0.069	1.416	0.981
	182.	3.196	4.950	-5.032	-3.718	-0.639	0.701	0.890
	214.	1.990	2.080	-1.322	-1.099	-0.972	0.709	0.949
	245.	0.000	0.250	0.098	-0.033	-0.257	-0.316	-0.272
	273.	0.000	8.740	-9.311	-8.640	-7.596	-6.644	-0.086
	301.	0.000	11.040	-10.394	-8.076	-2.248	2.979	4.880
	330.	0.000	7.290	-7.005	-7.145	-7.145	-7.145	-7.145
	361.	0.000	2.280	-2.104	-3.668	-9.095	-13.887	-13.887
	Total	25.785	62.670	-55.230	-48.185	-39.200	-29.748	-28.796

PLOT D

	Time (days)	Trans-piration	Infil-tration	depth(cm) 10	20	40	60	90
1981-1982	28.	0.000	0.790	0.324	-0.113	-2.709	-5.984	-8.209
	56.	2.090	4.470	-2.952	-2.153	-1.534	-1.820	-2.949
	84.	6.190	2.580	-1.001	0.555	1.714	1.031	-0.230
	112.	6.896	9.510	-7.679	-4.693	-1.690	-0.741	1.425
	140.	7.359	2.600	-1.051	1.174	2.867	2.105	-0.017
	168.	5.294	2.110	0.098	1.895	3.459	3.178	2.620
	196.	1.990	6.260	-4.642	-2.787	-0.469	0.430	1.470
	224.	0.020	3.910	-3.914	-4.277	-4.104	-3.360	0.270
	252.	0.000	6.830	-6.080	-5.455	-1.685	1.126	1.862
	291.	0.069	5.200	-5.496	-5.783	-7.839	-8.545	-8.545
	328.	0.020	2.840	-2.613	-2.794	-3.650	-5.805	-6.468
	357.	0.000	5.110	-3.949	-3.755	-3.238	-2.177	-1.738
	Total	29.927	52.210	-38.956	-28.186	-18.877	-20.562	-20.507
1982-1983	32.	0.000	1.250	0.058	-0.179	-1.400	-3.325	-6.588
	63.	2.300	3.410	-1.963	-1.219	-0.522	-0.530	-0.711
	91.	6.656	5.410	-3.318	-0.816	1.659	1.647	1.272
	123.	8.546	4.280	-3.208	-1.263	0.584	-0.277	-1.445
	154.	6.612	5.520	-3.073	0.115	3.614	3.833	4.060
	182.	5.376	1.040	-0.035	1.225	2.343	2.010	1.420
	211.	1.510	6.220	-4.463	-3.236	-1.311	-0.406	0.885
	243.	0.310	5.840	-5.223	-4.625	-4.051	-3.561	-2.233
	273.	0.170	6.530	-6.228	-5.760	-4.142	-0.976	2.151
	305.	0.610	6.440	-6.083	-5.833	-4.876	-4.748	-4.748
	337.	0.000	3.780	-3.439	-3.672	-4.824	-6.916	-6.916
	365.	0.000	1.450	-0.647	-0.609	-0.793	-1.155	-1.358
	Total	32.090	51.170	-37.621	-25.872	-13.722	-14.355	-14.210
1983-1984	34.	0.000	7.220	-5.466	-5.312	-4.573	-4.264	-4.264
	65.	1.020	9.660	-7.858	-6.999	-5.888	-4.558	-4.619
	91.	7.170	2.940	-1.591	0.063	0.527	-2.520	-4.429
	119.	9.170	3.290	-1.698	0.926	3.045	1.956	-0.853
	154.	8.425	4.290	-1.340	2.066	5.264	4.882	4.037
	182.	5.843	5.600	-4.242	-1.707	1.772	2.488	3.500
	214.	1.980	1.830	-0.588	0.025	0.635	0.555	0.315
	245.	0.000	0.270	0.288	0.278	0.204	0.132	0.032
	273.	0.000	9.980	-10.445	-9.945	8.342	-6.479	-1.737
	301.	0.000	11.230	-10.016	-8.363	-4.697	-2.065	-1.727
	330.	0.000	8.140	-8.048	-8.098	-8.098	-8.098	-8.098
	361.	0.000	2.780	-2.481	-3.545	-6.765	-8.532	-8.532
	Total	33.609	67.230	-53.485	-40.611	-26.917	-26.503	-26.375

Appendix 7. Annual fluxes of water and solutes in atmospheric input and in the soil of the four plots at different depths

Input values were derived in the same way as the underscored values in Table 41. Values are in mm for water, kmol/ha.yr for organic carbon (Corg) and SiO_2, and equivalent kmol/ha.yr (formerly: keq/ha.yr) for ionic solutes.

* refers to total dissolved carbon (organic + inorganic)

** refers to HCO_3-fluxes calculated from the difference in charge fluxes due to cations and anions (assuming organic anion fluxes to be negligible).

Plot A

1981/82	C_{org}	SiO_2	H^+	K^+	Na^+	Ca^{2+}	Mg^{2+}	Al^{3+}	NH_4^+	HCO_3^-	Cl^-	NO_3^-	SO_4^{2-}	water
input	8.48	–	0.74	0.03	0.80	0.30	0.18	–	3.30	–	0.79	0.84	3.42	514
10 cm	24.6	2.46	2.24	1.53	1.10	3.96	1.89	3.79	1.41	–	1.56	8.11	5.31	459
20 cm	11.9	2.27	1.41	0.88	0.95	2.91	1.63	4.58	0.27	–	1.23	8.32	3.48	376
40 cm	6.17	1.52	0.58	0.25	1.00	1.81	1.05	6.16	0.15	–	1.30	6.19	3.91	264
60 cm	3.42	1.19	0.09	0.03	0.96	4.70	1.81	1.07	0.03	–	1.47	3.94	3.60	182
90 cm	2.65	0.77	0.00	0.02	0.81	7.79	3.16	0.01	0.09	3.14**	0.97	4.73	3.04	186
1982/83														
input	6.27	–	0.67	0.04	0.87	0.40	0.21	–	2.61	–	0.94	0.91	2.84	519
10 cm	23.7*	2.05	1.68	1.10	1.04	4.13	1.53	3.67	0.99	–	1.80	8.65	3.75	453
20 cm	15.3*	1.90	1.10	0.75	0.87	2.28	0.93	4.03	0.29	–	1.40	5.76	2.82	366
40 cm	7.75*	1.59	0.49	0.32	1.23	2.21	1.34	7.17	0.10	–	1.14	8.45	3.45	256
60 cm	2.32*	0.96	0.04	0.06	0.93	4.65	1.83	0.52	-0.01	–	1.03	5.45	2.16	148
90 cm	3.82	0.39	0.00	0.01	0.97	8.64	3.06	0.30	0.05	2.02**	1.70	7.09	2.22	151
1983/84														
input	9.61	–	0.73	0.08	0.95	0.47	0.24	–	2.93	–	0.96	1.09	3.24	620
10 cm	25.3*	1.99	1.90	1.20	1.32	3.19	1.33	2.34	0.57	–	1.64	6.15	3.57	551
20 cm	13.8*	1.79	1.17	1.21	1.19	2.23	2.22	3.78	0.22	–	1.48	6.37	3.70	457
40 cm	6.00*	1.97	0.55	0.30	1.12	2.05	0.97	6.61	0.02	–	1.58	5.47	4.77	359
60 cm	4.11*	1.73	0.05	0.02	1.04	5.57	2.03	1.34	0.01	–	1.43	5.11	3.53	242
90 cm	4.14	1.01	0.00	0.03	1.20	11.6	3.39	0.07	0.01	6.07**	1.47	4.86	3.90	241

Plot B

1981/82	C_{org}	SiO_2	H^+	K^+	Na^+	Ca^{2+}	Mg^{2+}	Al^{3+}	NH_4^+	Cl^-	NO_3^-	SO_4^{2-}	water
input	8.44	–	0.72	0.03	0.80	0.30	0.18	–	3.18	0.79	0.95	3.18	517
10 cm	24.2	2.16	3.20	1.38	1.03	2.83	1.38	4.68	1.07	0.68	7.17	4.09	455
20 cm	17.2	1.38	2.29	1.26	0.73	2.89	1.37	4.28	0.61	1.05	6.97	3.99	376
40 cm	9.91	0.92	0.53	0.88	0.58	1.68	0.92	5.59	0.97	1.22	6.48	2.95	276
60 cm	4.29	0.49	0.18	0.34	0.32	0.75	0.45	3.06	0.16	0.69	2.67	1.79	166
90 cm	3.71	0.69	0.11	0.31	0.40	0.67	0.48	2.83	0.06	0.50	1.14	3.06	188
1982/83													
input	7.17	–	0.26	0.04	0.87	0.40	0.21	–	2.73	0.94	0.87	2.60	513
10 cm	25.2*	1.31	2.03	1.35	0.92	1.48	0.68	1.63	0.62	1.32	5.67	2.78	447
20 cm	16.2*	1.56	1.47	1.27	0.68	2.05	0.78	3.50	0.30	1.20	6.29	2.78	369
40 cm	10.5*	0.97	0.36	0.47	0.72	1.55	0.79	5.06	0.11	1.11	4.62	5.89	260
60 cm	3.32*	0.43	0.16	0.27	0.49	0.68	0.39	3.02	0.01	0.72	2.31	1.96	152
90 cm	2.85*	0.46	0.18	0.12	0.53	0.55	0.38	3.63	0.01	0.83	1.87	2.35	166
1983/84													
input	10.2	–	0.73	0.08	0.95	0.47	0.24	–	3.06	0.96	1.24	3.19	624
10 cm	28.3*	1.00	2.28	1.96	0.86	1.45	0.89	1.24	0.16	1.18	5.19	2.62	556
20 cm	19.5*	1.92	1.86	1.16	0.99	1.78	0.83	3.95	0.35	1.25	6.24	3.10	468
40 cm	13.2*	0.97	0.46	0.79	0.68	1.71	0.56	4.55	0.02	1.10	4.01	2.39	376
60 cm	5.92*	0.79	0.28	0.54	0.57	1.16	0.56	4.74	0.02	1.07	3.37	2.85	277
90 cm	5.45	0.75	0.30	0.21	0.71	0.96	0.48	4.98	0.02	1.23	2.92	3.53	264

Plot C

1981/82	C_{org}	SiO_2	H^+	K^+	Na^+	Ca^{2+}	Mg^{2+}	Al^{3+}	NH_4^+	Cl^-	NO_3^-	SO_4^{2-}	water
input	6.22	–	0.66	0.03	0.80	0.30	0.18	–	2.47	0.49	0.83	2.55	517
10 cm	23.8	2.68	0.91	0.70	1.10	1.78	1.01	3.80	0.79	1.27	3.43	4.64	467
20 cm	14.6	1.93	0.72	0.56	0.69	0.89	0.91	2.35	0.73	0.79	2.03	2.57	392
40 cm	8.77	1.45	0.26	0.49	0.66	1.21	1.34	2.82	0.15	1.11	3.12	2.43	310
60 cm	4.26	1.02	0.14	0.40	0.65	0.77	1.07	2.20	0.12	1.04	1.29	2.54	249
90 cm	3.06	0.89	0.03	0.17	0.55	0.75	0.93	3.06	0.12	0.70	1.05	2.73	247
1982/83													
input	5.89	–	0.28	0.04	0.87	0.40	0.21	–	2.20	0.94	0.75	2.55	519
10 cm	22.7	1.97	0.72	0.99	0.77	1.16	0.80	2.26	0.56	1.47	2.68	3.68	455
20 cm	15.2*	1.57	0.84	0.64	0.66	0.85	0.96	2.21	0.42	1.12	2.27	2.45	380
40 cm	9.16*	1.18	0.25	0.28	0.65	0.73	0.99	1.97	0.15	0.94	1.88	2.22	282
60 cm	3.68*	0.76	0.14	0.22	0.46	0.51	0.74	1.22	0.04	0.74	0.87	1.77	195
90 cm	3.65*	0.87	0.14	0.09	0.50	0.72	0.89	1.97	0.02	0.77	1.13	2.36	205
1983/84													
input	9.18	–	0.41	0.08	0.95	0.47	0.24	–	2.34	0.96	0.96	2.46	627
10 cm	28.8*	2.04	1.06	1.55	0.86	1.24	1.09	2.71	0.72	1.31	3.82	3.13	552
20 cm	18.9*	1.30	1.02	1.37	0.79	0.91	0.88	1.66	0.36	1.19	2.96	2.10	482
40 cm	9.08*	1.51	0.39	1.49	0.68	1.25	1.34	3.16	0.10	1.07	3.60	2.49	392
60 cm	3.48*	1.21	0.43	0.29	0.57	0.92	1.20	2.14	0.00	0.99	2.00	2.50	297
90 cm	4.88*	1.20	0.25	0.12	0.59	0.94	1.16	2.80	0.00	0.83	2.49	2.57	288

Plot D

1981/82	C_{org}	SiO_2	H^+	K^+	Na^+	Ca^{2+}	Mg^{2+}	Al^{3+}	NH_4^+	HCO_3^-	Cl^-	NO_3^-	SO_4^{2-}	water
input	10.7	–	1.37	0.03	0.80	0.30	0.18	–	3.72	–	0.79	0.98	4.34	522
10 cm	12.3	0.67	0.00	0.62	0.93	20.2	1.71	0.05	0.18	5.79	1.79	11.2	4.91	390
20 cm	7.57	0.61	0.00	0.14	0.84	16.2	1.40	0.04	0.07	4.76	1.56	8.71	3.66	282
40 cm	3.91	0.27	0.00	0.04	0.49	11.8	0.87	0.03	0.02	3.80	1.40	5.34	2.71	189
60 cm	3.92	0.33	0.00	0.02	0.60	9.55	1.93	0.20	0.02	2.99	0.21	5.05	2.89	206
90 cm	3.30	0.27	0.00	0.02	0.55	12.4	1.01	0.01	0.10	4.72	0.83	5.70	2.84	205
1982/83														
input	9.26	–	0.99	0.04	0.87	0.40	0.21	–	3.02	–	0.94	0.92	3.53	512
10 cm	15.3	0.64	0.00	0.37	1.01	14.4	1.45	0.08	0.24	4.45	2.27	6.79	4.04	376
20 cm	9.87	0.54	0.00	0.11	0.82	14.5	1.13	0.02	0.12	4.91	1.67	6.45	3.56	259
40 cm	3.79	0.22	0.00	0.02	0.52	8.62	0.56	0.02	0.05	2.28	0.90	4.13	2.48	137
60 cm	2.56	0.15	0.00	0.00	0.68	7.11	1.61	0.02	0.03	2.69	1.21	7.38	3.55	144
90 cm	2.02	0.12	0.00	0.01	0.44	9.63	0.88	0.01	0.01	3.55	1.03	4.62	1.78	142
1983/84														
input	11.9	–	1.00	0.08	0.95	0.47	0.24	–	3.65	–	0.96	1.17	4.15	672
10 cm	18.7	0.68	0.00	0.93	1.36	19.2	1.42	0.06	0.03	6.05	2.01	10.6	4.34	535
20 cm	10.3	0.67	0.00	0.17	1.13	14.9	1.17	0.06	0.02	8.12	1.69	4.50	3.14	406
40 cm	4.54	0.33	0.00	0.04	0.92	15.5	1.02	0.01	0.00	5.94	1.92	6.28	3.35	269
60 cm	5.67	0.37	0.00	0.01	1.07	15.1	2.09	0.00	0.01	4.81	1.55	7.76	4.16	265
90 cm	4.87	0.25	0.00	0.03	0.84	18.0	1.60	0.02	0.00	6.41	1,74	8.33	4.01	264